BIO-OBJECTS

Theory, Technology and Society

Series Editor: Ross Abbinnett, University of Birmingham, UK

Theory, Technology and Society presents the latest work in social, cultural and political theory, which considers the impact of new technologies on social, economic and political relationships. Central to the series are the elucidation of new theories of the humanity-technology relationship, the ethical implications of techno-scientific innovation, and the identification of unforeseen effects which are emerging from the techno-scientific organization of society.

With particular interest in questions of gender relations, the body, virtuality, penality, work, aesthetics, urban space, surveillance, governance and the environment, the series encourages work that seeks to determine the nature of the social consequences that have followed the deployment of new technologies, investigate the increasingly complex relationship between 'the human' and 'the technological', or addresses the ethical and political questions arising from the constant transformation and manipulation of humanity.

Other titles in this series

Decentering Biotechnology
Assemblages Built and Assemblages Masked
Michael S. Carolan
ISBN 978 1 4094 1005 8

The Genome Incorporated
Constructing Biodigital Identity
Kate O'Riordan
ISBN 978 0 7546 7851 9

Technology and Medical Practice
Blood, Guts and Machines
Edited by Ericka Johnson and Boel Berner
ISBN 978 0 7546 7836 6

Contested Categories
Life Sciences in Society
Edited by Susanne Bauer and Ayo Wahlberg
ISBN 978 0 7546 7618 8

Bio-Objects
Life in the 21st Century

Edited by

NIKI VERMEULEN
University of Vienna, Austria

SAKARI TAMMINEN
University of Helsinki, Finland

ANDREW WEBSTER
University of York, UK

Routledge
Taylor & Francis Group

LONDON AND NEW YORK

First published 2012 by Ashgate Publishing

Published 2016 by Routledge
2 Park Square, Milton Park, Abingdon, Oxon OX14 4RN
711 Third Avenue, New York, NY 10017, USA

First issued in paperback 2017

Routledge is an imprint of the Taylor & Francis Group, an informa business

British Library Cataloguing in Publication Data
Bio-objects : life in the 21st century. -- (Theory,
    technology and society)
    1. Life (Biology)--Philosophy. 2. Genetic engineering.
    3. Genetic engineering--Moral and ethical aspects.
    4. Transgenic organisms. 5. Human reproductive technology.
    6. Human reproductive technology--Moral and ethical
    aspects.
    I. Series II. Vermeulen, Niki. III. Tamminen, Sakari.
    IV. Webster, Andrew, 1951-
    660.6'5-dc22

Library of Congress Cataloging-in-Publication Data
Vermeulen, Niki.
    Bio-objects : life in the 21st century / by Niki Vermeulen, Sakari
Tamminen, and Andrew Webster.
        p. cm. -- (Theory, technology, and society)
    Includes bibliographical references and index.
    ISBN 978-1-4094-1178-9 (hbk)
    1. Life sciences--Social aspects. 2. Transgenic animals.
    I. Tamminen, Sakari. II. Webster, Andrew, 1951- III. Title.

QH333.B48 2011
508--dc23

                                                              2011040150

ISBN 13: 978-1-138-30650-9 (pbk)
ISBN 13: 978-1-4094-1178-9 (hbk)

Contents

List of Figures and Tables

Figures

Tables

Notes on Contributors

Bettina Bock von Wülfingen is a biologist with a PhD in public health, now Assistant Professor at the Institute for Cultural History and Theory at the Humboldt-Universität zu Berlin. Her work revolves around epistemologies in Life Sciences, always applying a historical perspective. Her current, second book project deals with methodology, more explicitly with the interrelation of instruments, background beliefs and models in heredity and reproductive sciences around 1900 and 2000. She did her doctoral thesis with a discourse analysis on the health and illness notion in discourses on new reproductive and genetic technologies. During her PhD, she received different international visiting fellowships within Science and Technology Studies, working *inter alia* at the IAS-STS in Graz/Austria, the SATSU in York and the BIOS in London.

Nik Brown is Reader in Sociology and co-director of the Science and Technology Unit (SATSU) at the University of York. His research interests focus on culturally intriguing developments in the biosciences such as cloning, transpecies transplantation, hybrids, chimeras, stem cells, and biobanking. He is interested in the social management of the boundaries between life and death, the human and the animal, the biologically mundane and the exotic, the public and the private. He is particularly interested in the politics, regulation and governance of novel biological developments and reproduction. He has also written extensively on the sociology of hope, expectations and futurity.

Conor M.W. Douglas is a post-doctoral Researcher in both the Section Community Genetics at the Vrije University (Amsterdam) medical centre, and in the Technology Assessment group at the Rathenau Institute in The Hague. Originally trained as a sociologist and then in science and technology studies (STS), his research interests are located in the interactions and co-production between biomedical sciences and society. This general interest has resulted in empirical research in patient and 'user' involvement in biomedical innovation processes including pharmacogenetics, biobanking, bioinformatics, synthetic biology, meta-genomics and bioremediation, and translational science more generally.

Lena Eriksson is Assistant Professor at the Department of Philosophy, Linguistics and Theory of Science, University of Gothenburg, Sweden. Her main research interests are effects of taxonomies and standards in science and medicine; which she explored in an ethnographic study of international standardisation efforts in human embryonic stem cell research. Her two most current research projects examine the understanding and management of pain in varying health care contexts and the production of a new standardised decision-making tool for Swedish health care practitioners. She holds a PhD from Cardiff University and has also worked in the Science and Technology Studies Unit (SATSU) at the University of York.

Janus Hansen is Associate Professor of Sociology at the Department of Business and Politics, Copenhagen Business School. He holds a PhD from the European University Institute in Florence (2005). Prior to his posting at CBS he has taught and conducted research at several other universities in Denmark. His research interests include sociology of science and technology, sociology of risk, social theory and comparative sociology. He is particularly interested in the comparative studies of public engagement with science and technology and the interplay between science and politics. Recent publications include a research monograph entitled 'Biotechnology and Public Engagement in Europe' (2010).

Tora Holmberg holds a PhD in sociology and is an Associate Professor at the Institute for Housing and Urban Research, Uppsala University, Sweden. Her focus lies at the intersection of animal studies, STS (science and technology studies) and feminist science studies, researching human-animal relations in bio-technology as well as in the urban setting.

Ine Van Hoyweghen is assistant professor at the Department of Health, Ethics & Society (HES), Maastricht University. She holds a PhD in Social Sciences from the Catholic University of Leuven (Belgium), which was published as a book, *Risks in the Making. Travels in Life Insurance and Genetics* (Amsterdam University Press, 2007). Her work focuses on the social implications of biomedicine and biomedical technologies. She is interested in the hybrid processes of classification in biomedicine, states and markets in producing identities and social relations. Her empirical research covers the domain of European life and health insurance. She recently guest edited a special issue of *New Genetics & Society* (2010, 29[4]) on the *mattering* of solidarity in private and social arrangements. She is currently developing an account of 'genetic discrimination' in European insurance.

Malin Ideland is Associate Professor in European Ethnology and Senior Lecturer in Educational Science at Malmö University. Her main research and teaching areas are 1) cultural perspectives on biotechnologies and 2) science education for citizenship and sustainable development. In her research on cultural perspectives on biotechnologies, she has studied media reports on gene technology, debates on embryonic stem cell-techniques, parents decision-making on MMR vaccines and dilemmas concerning transgenic animals.

Ingrid Metzler is a PhD student at the Department of Political Science at the University of Vienna, Austria, and a junior researcher at the Life-Science-Governance Research Platform at the same university. She has worked on several research projects exploring the governance of biomedicine in Europe. With a background in political science, she is particularly interested in the co-construction of bio-objects and regimes of governance.

Nete Schwennesen is a sociologist, and holds a PhD from the Center for Medical Science and Technology Studies at the Institute of Public Health, University of Copenhagen. For her PhD she studied processes of knowledge production and decision making in the context of first trimester prenatal risk assessment, on the basis of ethnographic material. She has a Masters degree in contemporary sociology from Lancaster University, UK and was a Marie Curie fellow at Science and Technology Studies Unit (SATSU), University of York, UK in 2005. Her research interests centres on the co-production of agency and health care technologies, new reproductive technologies, feminist science studies and issues of reproductive choice. Currently she works as a researcher at Steno Diabetes Health Promotion Center in Copenhagen, where she studies programmes of patient education from a social setting perspective.

Sakari Tamminen gained his PhD from the University of Helsinki, Finland, where he graduated in social psychology and social anthropology. His research interests lie in the exploration of human-nonhuman boundaries, new genetic technologies and the insights these provide for rethinking social theory. He also holds a Licentiate of Technology degree in usability and user-centered design from Aalto University where he acted as university lecturer for a number of years. He has also worked as a Marie Curie Research Fellow at the Science and Technology Studies Unit (SATSU) at the University of York, UK (2004-2005) and as a research fellow at the Institute for Advanced Studies on Science, Technology and Society (IAS-STS) in Graz, Austria, in 2007. He is currently interested in efforts to digitalise and recreate life in all of its shapes and is an Academy of Finland post-doctoral researcher at the University of Helsinki, Finland.

Aaro Tupasela works as a researcher in the Department of Social Research at the University of Helsinki and is a member of the Helsinki Institute of Science and Technology Studies (HIST). His areas of interest include the sociology of science, the sociology of knowledge, public understanding of science and bioethics. He also works as a Finnish representative on the Nordic Committee on Bioethics and serves as a board member of the European Sociological Association's (ESA) research network on science and technology studies (SSTNET).

Niki Vermeulen is a post-doctoral research fellow in the Department of Social Studies of Science at the University of Vienna, Austria, and is consultant for Technopolis Group. Niki holds a PhD in Science and Technology Studies from Maastricht University, The Netherlands, and worked for the Executive Board of Maastricht University, the Scientific Council of the Royal Netherlands Embassy in Washington DC and the Netherlands Scientific Council for Government Policy (WRR). Her research interests concern transformations in research, and science and innovation policy. She is particularly interested in scientific collaboration, focusing on the life sciences and looking into both ecology and (post-)genomics research.

Andrew Webster is Professor in the Sociology of Science and Technology and Director of the Science and Technology Studies Unit (SATSU) at the University of York, UK. He was Director of the £5m ESRC/MRC Innovative Health Technologies Programme, and was national coordinator the ESRC's £3.5m Stem Cells Initiative (2005-2009), and member of the Royal Society's Expert Working Group on Health Informatics. He is currently undertaking externally funded research on stem cells as well as the implementation of pharmacogenetics into clinical practice, and is coordinating a European project on Regenerative Medicine (REMEDiE). He is co-editor of the *Health Technology and Society* Series (Palgrave Macmillan). His most recent book is *Health, Technology and Society: A Sociological Critique* (Palgrave Macmillan, 2007). He was elected a Fellow of the Academy of the Social Sciences in 2006.

Ragna Zeiss is Assistant Professor at the Department of Technology and Society Studies, Maastricht University, the Netherlands. She obtained her PhD from the Science and Technology Studies Unit (SATSU) at the Department of Sociology, University of York, UK. As a post-doctoral researcher, she worked on the project 'Rethinking Political Judgement and Science-Based Expertise: Boundary Work at the Science/Politics Nexus of Dutch Knowledge Institutes' at the VU University Amsterdam. Her main research interests concern water, sanitation, and the environment; standardisation, regulation and governance; knowledge brokerage and boundary work; risk governance; and sustainability. She is currently involved in European and national (research) projects on brokering sustainable sanitation; nanotechnologies, risk governance, and development in India, Kenya and the Netherlands; bio-objects and their boundaries; and science, ethics, and technological responsibility in developing and emerging countries.

Acknowledgements

The origin of this book lies in York, United Kingdom, in the Science and Technology Studies Unit (SATSU) in the Department of Sociology of the University of York, that was a Marie Curie training site in the period 2002-2005. With the title 'New Genetics/New Society? Integrating Science, Society and Policy', it attracted 15 European doctoral students working on recent developments in bioscience and society. Such was the success of the programme the idea of continuing our collaboration beyond the formal end of the EC contract was welcomed by all: we now have a well-established group that has met for annual meetings (York 2007; Geneva 2008; Uppsala 2009; Graz 2010). In these meetings, we developed our collective work on 'bio-objects' which resulted in this book. In 2010 the network has been granted a COST Action entitled 'Bio-objects and their Boundaries: Governing Matters at the Intersection of Society, Politics, and Science'. This means that our network is expanding and we will be continuing working together in the period 2010-2014 and beyond to further develop the bio-objects concept.

First and foremost, we would like to acknowledge all the members of SATSU that made our stay in York such a pleasant intellectual experience. We are also grateful for the support that our network has received from several other organisations: The Brocher Foundation in Geneva, Switzerland, the Centre for Gender Research, Uppsala University, Sweden, Riksbankens Jubileumsfond, the European Association for the Study of Science and Technology (EASST), the Society for the Social Studies of Science (4S), the Institute for Advanced Studies on Science, Technology and Society (IAS-STS) in Graz, Austria, the Faculty of Arts and Social Sciences of Maastricht University, The Netherlands, and the European Cooperation in Science and Technology (COST). Finally we want to thank Neil Jordan and other staff from Ashgate who provided us with the necessary support for the publication of the book, as well as the sub-editors of the three parts: Conor M.W. Douglas, Aaro Tupasela, Bettina Bock von Wülfingen and Ingrid Metzler.

Introduction
Bio-Objects: Exploring the Boundaries of Life

Andrew Webster

Biotechnologies and new biological artefacts are currently disrupting the conventional boundaries and identities of biological forms – whether these be of a human, animal, plant or synthetic nature. Indeed these discrete forms of life may well be brought together, hybridised, through developments in the biological sciences, such as in tissue engineering. At the same time, these new life-forms – such as pluripotent stem cells outside of bodies, synthetic biology or genetically modified organisms – create new clinical (Mason 2007) and commercial (Waldby 2006) possibilities as well as regulatory demands (Brown and Kraft 2006).

This book conceptualises these different life forms as what we shall call 'bio-objects'. Bio-objects play a crucial role in the 21st century in which increasing knowledge of life and its components are fundamentally transforming what life means and where its boundaries lie: as Thacker (2010) says, ' …it seems that life is everywhere at stake, and yet it is nowhere the same' (ix). New developments in the biosciences – especially the molecularisation of life – and their influence on health care and other aspects of our society have been explored in a diverse body of literature, discussing the ethical (e.g. Holm 2005), legal (Johnston and Kaye 2004) and social (Rose 2006) implications of these new developments. In the following chapters we will draw on an extensive body of work to ground our own material which is primarily though not exclusively underpinned by and seeks to make a contribution to science and technology studies (STS). In this introductory chapter, we sketch out some of the basic ideas we will expand on in later chapters about the meaning of the bio-object, give some indication of how this work relates to existing literature and outline the structure of the book as a whole.

The meaning of bio-objects and bio-objectification

We want to suggest that the term 'bio-object' is a useful conceptual device or heuristic to refer to socio-technical phenomena where we see a new mixture of relations to life or to which 'life' is attributed. As a consequence of these novel relations, the boundaries between human and animal, organic and nonorganic, living and the suspension of living (and the meaning of death itself), are questioned and destabilised, though sometimes can be re-established or re-confirmed. This dynamic process, and the possibility of moving backwards and forwards between different life statuses suggests that there is no once-and-for-all list of bio-objects, a sort of bio-object catalogue, made up of life forms that have specific properties or essential characteristics. Instead, we want to argue – and show in this book – that it is more useful to focus on the *process* of bio-objectification, that is to say, how different life forms are created and are given life, and perhaps, multiple lives. Aborted fetal tissue is a simple example of this process that is deemed, as such, to be waste matter and dead, yet can be re-vitalised as source material for stem cell lines. Here, as elsewhere, life is in constant interplay with novel techniques aiming at re-routing, diversifying, collecting and commodifying the vital processes that 'life' consists of.

A number of deep going philosophical questions are raised by the emergence of technologically enacted vital materiality – for example cell life imitated by silicon chips or deep frozen animal reproductive material – as they contest our contemporary understanding of 'life' and its boundaries. What could be said about organic life does not apply easily to forms of life that are technoscientifically created and sustained – that might be regarded as 'creatures' '…positioned somewhere between the normal and pathological' (Thacker 2010: 97). At the same time these novel forms of vitality also highlight the materiality of vitality in surprising ways. Bio-objectification clearly requires working with the materiality of the biological but not all biological objects – say the gene – can be seen as bio-objects, even though they may become the basis for them.

To be more precise, particular instances or collections of matter *become vital* in/to different practices and knowledge regimes, which at the same token *make life matter*. Historically speaking, the knowledge regime of biology, hence also 'biological objects', are of recent invention as Foucault (1970) and other historians of life sciences have famously argued. What we are currently witnessing, however, is that 'life' as an *object* of research, intervention and innovation is increasingly represented through an idiom of science and its unquestionable regime of truth – both in academic literature as well as in science communication aimed at different publics. A number of academic contemporary writers (e.g. Rose 2006) even claim that humanity has crossed a threshold where life has become totally manageable through its 'molecularisation/genetisation' as a result of a revolutionary co-development of scientific understanding of life's basic components and the technologies capable of representing and modifying them (e.g. molecular structure and genetic engineering). In short, the argument is that the total 'objectification' of life has happened through breakthroughs in scientific knowledge and related representation and intervention techniques.

While the 'molecularisation/genetisation thesis' certainly holds true for certain epistemic cultures (Knorr-Cetina 1999) of our technoscientific modernity (such as bio-medical research communities), the relevant collection of things related of life -processes, materialities and social/political issues – are not so clearly cut in other communities or for all actors involved. In fact, the argument of molecularisation of life dangerously flirts with the essentialisation of life as molecules and DNA, the reduction of the meaning of living and materialities of life to an essentially contemporary biological thinking. What we argue in this collection of articles is that in order to resist the temptation of this reductionist thesis we have to look in detail to the way in which life is *made an object* in different settings – both in and outside of the current truth regime of the contemporary biosciences. Instead, life – and different forms of life that are created – become an object of knowing, representing and intervening in a myriad of ways not reducible to the molecurisation thesis. Rather than reduction, we see hybridisation everywhere, the ongoing 'border crossings' (Haraway 1997: 60) made possible by the biosciences and by the socio-technical cultures that enable their movement.

Given this state of affairs, we expect to see bio-objects as being characterised as having considerable fluidity and mobility across different socio-technical domains or arenas. This means that a bio-object, associated with, say, biomedical research, may find its way into the food system or the environment, become part of a repository and new medium of technical innovation (as in biobanks or cord blood banks), and have multiple or even contrasting cultural meanings as it circulates between different sectors or networks of society. At the same time, new regulatory boundaries are developed for what human and non-human material can and cannot be legitimately traded as bio-objects (for example, oocytes and embryos) (Hauskeller, Bender and Manzei 2005).

Much of the recent literature in STS that has examined the issues central to this collection has focused on the field of biomedicine and particular areas within it, such as genetics/genomics (Atkinson et al. 2009), the tissue economy (Waldby and Mitchell 2006; Sunder Rajan 2006) and through its commercialisation, the exploitation of reproductive and clinical labour (Cooper 2008). The biomedical domain is clearly one where matters of life and what matters, and definitions of normal and abnormal life, provide a rich ground for STS scholars to explore the life-technology relationship. Such work shows that what is deemed to be medically normal is highly dependent on what technologies exist and how these – through greater diagnostic power for example – define the boundaries of the normal and the pathological (Canguilhem 1998; Lock and Nguyen 2010); Rose (2007) calls these – echoing Foucault – 'technologies of optimisation'. The book builds on this corpus as well as recent work on biological citizenship (Rose 2006), the contestation over categories of the biological (witnessed e.g. in the emergence of 'biomedia' as Thacker, 2005, argues, [see also Bauer and Wahlberg, 2009]) and the precariousness of new life (see, e.g. Mesman 2008, and her study of premature birth).

Beyond the biomedical domain, work by STS scholars on the role of the taxonomic sciences (e.g. Waterton et al. 2010) shows how taxonomy itself as the classification and so in effect disciplining of life (and indeed, of scientific communities) changes over time as new techniques are developed, such as genetic determination of species boundaries. More importantly for our purposes, this work also makes clear how taxonomies are best seen as epistemic assemblages, that is, always relational and subject to change as to what 'counts' as 'in' and 'out'. Related work on standardisation (e.g. Bowker and Star 2000; Eriksson and Webster, 2009) shows how classification procedures not only seek to order and bring closure to – 'sort out' – life forms (such as 'race') and behaviour (such as the allocation of illness categories) but also carry moral dimensions and presumed hierarchies of life. Bio-objects can be closed down or opened up as they fall within or move across classificatory regimes – such as in insurance systems (e.g. Van Hoyweghen, this book) or international databases to do with crime, biodata, identity, or indeed at a mundane level, as to what goes into what recycling box (Neyland and Woolgar, 2010). More generally, they can be directly linked to the reorganisation of scientific disciplines themselves, as Vermeulen shows (Chapter 11, this book).

Classification is, in one sense, a form of governance inasmuch as it provides one of the bases on which the regulation of life can occur, establishing boundaries of responsibility, inclusion and exclusion, and accountability. Governance forms an important theme in the book to encompass both formal state regulation and the 'soft law' that monitors and steers bioscience according to culturally derived normativities (Gottweis et al. 2008). More recently, Faulkner's (2009) work on 'governation' provides a conceptual framework that links governance with innovation, and shows how the two are co-constructed (Brown and Michael 2004).

The governance and materiality of bio-objects: Making matters of vital concern

In this collection we see governance as having two inter-related processes: governance of the bio-object, precisely through governance being generative of the bio-object qua *object* (as Metzler shows Chapter 10, this book), and thereby governance through the bio-object. So governance and the other core theme of the book – materiality – come together, for governance also points to the social requirements associated with the traceability of bio-objects and their moral status within different domains. The play of governance/materiality relations can vary considerably, such as enabling hybridity or closing it down. For example, as Brown argues later in this book (Chapter 5), UK legislation on interspecies embryos has successfully enshrined 'the novel unpredictability of hybrid biology in legal statute.' In contrast, during the Bush administration in the US, university researchers were barred from storing human embryonic stem cells from unapproved lines in lab refrigeration units paid for with NIH (National Institutes of Health) grant money and had to have designated 'private' and 'federally-funded' fridges for different cell lines. The two sets of lines as material objects carried quite different moral statuses, which were then reflected in two discrete fridges, which, as a result, were themselves of distinct material and moral status.

Paralleling the question of the material and moral traceability of bio-objects is the question of the tradability that they have. Indeed these two are typically closely tied together: for example, a novel biomedical therapy or device becomes marketable, tradable, only when it has secured a licence to be marketed through approval by regulatory agencies, such as the European Medicine Agency, and meets the requirements set down in various 'directives', such as the Advanced Therapy and Medicinal Product Directive (ATMP). The ATMP gives authorisation for a product to be placed on the market.

Outside of the commercial market for therapies or products, bio-objects – such as transgenic mice – become tradable across labs where they are deemed to conform to the standards of a high quality 'pure' transgenic mouse, but simultaneously framed as ordinary lab mice, as one of the chapters (Holmberg and Ideland) in the book explores. Tradability and traceability are in turn linked, for bio-objects move across and are commodified within different trading zones that are more, or less, open. Genetically-modified crops, for example, may (as in the US) or may not (as in the UK) be grown commercially, are subject to regulatory and trading constraints and various forms of containment, including biocontainment strategies which uses the genetic internalisation of regulation to overcome the failings of governance regimes in managing the problem of policing. In one sense, plants become self-policing. A further issue is where to draw the boundary lines between what is, and is not, subject to regulatory control: in the European Union, while food from cloned animals cannot be sold without authorisation as they fall under the terms of the 'novel food directive', there is some uncertainty as to whether their offspring are too (Mahony 2010).

These different themes – of the attribution and malleability of life, its materiality, governance and movement across different trading and regulatory spaces – are key to our understanding of the bio-object and the conceptual model we have of the bio-objectification process, and figure in different ways in the chapters that make up this book. They all share too a particular methodological approach that draws on a number of STS strategies, and these are sketched out below.

Methodological approach

As we are concerned in this book to understand how bio-objects come to be and the lives that they take-on we are interested, as noted above, in bio-objectification as a process. This means that the methodological approach – the logic of our inquiry – is to understand the interplay of material and epistemic dynamics in each of our cases, and to look at the ways in which boundaries of life are disrupted or conserved, are disentangled from, or entangled with, other forms of life.

We regard bio-objectification as an emergent process inasmuch as it is shot through with uncertainty and unknowns (Webster and Eriksson 2009). We also recognise (as Pickering) that materiality has agency, or performativity, inasmuch as the object – as we saw above with GM crops – can act at a distance in ordering social (farming) practices. This performativity is, as Pickering says, a 'dance of agency', but one we would characterise as reflecting an unstable ontology, an ongoing process rather than a stable form of being – so, our chapter on transgenic mice describes them in this regard as 'boundary crawlers'. As such, bio-objects contest the boundary lines between entities we have been accustomed to take for granted, as existing by themselves and for themselves.

While the authors all see the importance of understanding the circumstances within which the bio-objectification process takes place. Despite the importance then of understanding the local setting through which the process occurs, this does not mean that we seek to explain the meaning of the bio-object as being in some way read from a given context or stable across different contexts, for we share the STS approach that the material/context relation is plastic and is given meaning different socio-technical assemblages (Latour 1993). Our task then is to see whether and how such assemblages are more, or less, robust through time and space, whether they hold their shape or not (Law 2004), and the affordances they allow, which are in turn dependent on the interplay of the materiality of the bio-object, cultural knowledge and practices deployed in using it and whether it is available to all, or not and why (Bloomfield et al. 2010).

With this book we want to trace a variety of contemporary bio-objects in their emergence, stabilisation and circulation across different societies and different histories. Here, we draw on diverse empirical investigations that provide new ways of thinking about how novel bio-objects enter our contemporary life and societies in the 21st century. The different chapters report on a diverse range of methods – from ethnography, interviews, documentary sources, analysis of grey literature and so on – to tell their stories. The particular bio-objects that appear in the book reflect not merely the discrete research interests of the separate authors, but also exemplify the range of bio-object and bio-objectification that we believe characterise the contemporary bio-world. All address issues associated with hybridity and/or transgeneity and offer insight into the relation between the material and the epistemic, as boundaries of bio-objects are built or reconfigured. We share Barry's (2001) perspective that we should see the novelty of bio-objects covered here not in terms of some intrinsic property that makes them in some way radically new, but rather what, through the bio-objectification process, novel socio-technical (including political) relations are made possible – are opened up or afforded – by this process.

Taking this argument a little further, the bio-objectification process can open up new and maintain existing affordances: the bio-objectification process associated with IVF enables both the reproduction of children but also the production of supernumerary embryos for stem cell research: these embryos occupy a very different spatial and temporal universe associated with the tissue economy. Those involved – patients, carers, clinicians, and regulators – confront both of these continuities and discontinuities though not necessarily at the *same* time and not all are asked to bear the responsibility or mobilise social relations and networks to cope with the different challenges they pose.

The substantive focus of the book draws on this broad methodological orientation and includes analyses of the boundaries of life and living associated cloned animals, embryos, cybrids, genetic resources, as well as the apparently mundane world of water and the everyday clinical research subject. We have also sought to cover bio-objectification that explicitly includes consideration of organisational and institutional processes, as in the construction of biobanks, and the ways in which governance 'works' at this meta-level. We have also included material that covers activity across very different temporal-spatial locations from the lab through to the global research network of 'big science', and how we can see – for example in regard to national policy and legislation relating to IVF and embryos – the co-production of the state as a regulatory actor and the bio-object itself. Our approach then is to provide a broad range of examples of the making of the bio-object on a shared analytical framework. This is reflected in the structure of the book itself, which has been organised around three themes, as we now go on to describe.

Structure of the book

The conceptual territory sketched out above is explored further in the following chapters that have been organised as follows.

Part 1 examines the changing *boundaries* of the human, nonhuman and society as a result of the emergence of new bio-objects. Here we encounter transgenic beings and practices that constitute them, artificial bodies transgressing the boundaries of life and death and the boundary work that allows a highly manufactured bio-object – water – to be configured as 'pure' and 'natural'. We also explore how the meaning of 'the patient', a central site for the deployment of various bio-objects (such as the stem cell) is itself open to multiple configurations, not as a patient but as a 'research subject'. Common to these chapters is a focus on the ontological provisionality of the bio-objects, the meaning of 'the real', the 'natural', the 'other'.

Part 2 concentrates on the *governance* of new bio-objects and the social regulations involved in the boundary shifts that they bring about. We see different aspects, covering the governance of hereditary diseases, the regulation of cloned animal products, different forms of public engagement with GMOs, governance regimes around tissue banks and the paradoxes involved in the governance of life-not-yet-born through new testing technologies. An important theme underpinning this work is the ways in which governance often implies the governance of material infrastructures which in effect solidify the social regulations ob bio-objects.

Part 3 discusses the *new social, economic and political relations* that constitute and are constituted by these changes. It starts with a critical examination of the 'newness' of these relations looking specifically at practices in a very traditional area – insurance – and how these are changing in the molecular age. Then it examines ways in which different government regimes are deploying new bio-objects in both collective cultures and within highly individualised, neo-liberal societies. This empirical section ends with an analysis of the emergence of global research networks around new bio-objects that both create and are created by the objects themselves.

In drawing on the contributions from within STS and wider social/political theory, some of which have been noted above, we hope that the book provides a broad-ranging exploration of the meaning of life in the 21st century, and that through the concept of the bio-object provides a series of conceptual tools linked in particular to notions of 'geneity', 'hybridity' and 'generative relations' through which we can interrogate the continuities, breaks, and implications of different 'life forms'. These implications are shot through with uncertainty not least as they are often framed by narratives of promise and breakthrough accompanied by both hype and new anxieties. To that extent, the various chapters in the book offer a critical engagement with bio-objectification and various authors are prepared to adopt a normative perspective in building their critique. Such a critique has academic value in helping to provide fresh insight into the boundaries of life debates that have characterised the STS commentary on the biosciences that we have seen over recent years. But it also should, in the longer term, have policy relevance.

There are a series of policy challenges that lie ahead that result directly from the bio-objectification process not only within but between countries, not least because of the mobilisation of bio-objects through international trading zones. For example, there is a need to understand how to oversee and regulate the international exchange of tissue samples between European and US stem cell banks given that the sourcing and procedures vary so much. Is it important that we see variability in the bio-objectification processes across these banks; how and in what optimal form should these processes be standardised? In the longer term there is a need to understand how the resources such public banks hold will be open to private exploitation through commercialisation and so enter a different trading zone, an increasingly likely process as they move towards budgets based on self-financing through the gradual reduction of state support. These and other policy issues (such as the meaning of 'the embryo') figure throughout the book.

Finally, our own work is – to use an appropriate metaphor – embryonic inasmuch as this book is part of a wider research endeavour which involves modeling the bio-objectification process in other domains not included here (such as synthetic biology). However, we believe that the core features of this model which I have tried to sketch out above provide a good basis for this future work.

References

Atkinson, P. et al. 2009. *Handbook of Genetics and Society*. London: Routledge.

Barry, A. 2001. *Political Machines: Governing a Technological Society*. London: Athlone Press.

Bauer, S. and Wahlberg, A. (eds) 2009. *Contested Categories: Life Sciences in Society*. Aldershot: Ashgate.

Bloomfield, B. et al. 2010. Bodies, technologies and action possibilities: When is an affordance? *Sociology*, 44(3), 415-34.

Bowker, G. and Star, S.L. 2000. *Sorting Things Out: Classification and its Consequences.* Cambridge, MA: MIT Press.

Brown, N. and Kraft, A. 2006. Blood ties: Banking the stem cell promise. *Technology Analysis and Strategic Management,* 18(3/4), 313-27.

Brown N. and Michael, M. 2004. Risky creatures: Institutional species boundary change in biotechnology regulation. *Health, Risk and Society,* 6, 207-22.

Canguilhem, G. 2007. *The Normal and the Pathological.* New York: Zone Books.

Cooper, M. 2008. *Life as Surplus: Biotechnology and Capitalism in the Neoliberal Era.* Washington: Washington University Press.

Faulkner, A. 2009. *Medical Technology into Health Care and Society.* Basingstoke: Palgrave Macmillan.

Foucault, M. 1970. *The Order of Things.* London: Tavistock.

Gottweiss, H., Salter, B. and Waldby, C. 2008. *The Global Politics of Human Embryonic Stem Cell Research.* Basingstoke: Palgrave Macmillan.

Haraway, D. 1997. *Modest _Witness @ Second Millennium.FemaleMan _Meets _Oncomouse.* New York and London: Routledge.

Hauskeller, C., Bender, W. and Manzei, A. 2005. *Crossing Borders.* Münster: Agenda Verlag.

Holm, S. 2005. Embryonic stem cell research and the moral status of human embryos. *Reproductive Biomedicine Online,* 10 (Sup. 1) 63-7.

Johnston, C. and Kaye, J. 2004. Does the UK Biobank have a Legal Obligation to Feedback Individual Findings to Participants? *Medical Law Review,* 12(3), 239-67.

Knorr-Cetina, K. 1999. *Epistemic Cultures: How the Sciences Make Knowledge.* Cambridge, MA: Harvard University Press.

Latour, B. 1993. *We Have Never Been Modern.* Cambridge, MA: Harvard University Press.

Law, J. 2004. *Organizing Modernity.* Oxford: Blackwell.

Lock, M. and Nguyen, V-K. 2010. *An Anthropology of Biomedicine.* London: Wiley-Blackwell.

Mahony, H. 2010. Milk from cloned cow offspring exposes gap in EU food law. Available at: www.EU.observer.com [accessed 4 August 2010].

Mason, C. 2007. Regenerative Medicine 2.0. *RegenMed,* 2(1), 11-18.

Mesman, J. 2008. *Uncertainty in Medical Innovation: Experienced Pioneers in Neonatal Care.* Health, Technology and Society Series. Basingstoke: Palgrave Macmillan.

Neyland, D. and Woolgar, S. 2010. Traffic Safety and Control, in *Globalization in Practice,* edited by A. Tickle, N. Thrift and S. Woolgar. Oxford: Oxford University Press.

Rose, N. 2006. *Politics of Life Itself: Biomedicine, Power and Subjectivity in the Twenty-first Century.* Princeton, NJ: Princeton University Press.

Sunder Rajan, K. 2006. *Biocapital: The Constitution of Postgenomic Life.* Durham, NC: Duke University Press.

Thacker, E. 2005. *The Global Genome: Biotechnology, Politics, and Culture*. Cambridge, MA: MIT Press.

Thacker, E. 2010. *After Life*. Chicago, IL: University of Chicago Press.

Waldby, C. 2006. Umbilical Cord Blood: From Social Gift to Venture Capital, *BioSocieties*, 1(1), 55-70.

Waldby, C. and Mitchell, R. 2006. *Tissue Economies: Blood, Organs and Cell Lines in Late Capitalism*. Durham, NC: Duke University Press.

Waterton, C. et al. 2010. *Barcoding Nature: Shifting Taxonomic Practices in an Age of Biodiversity Loss*. London: Routledge.

Webster, A. and Eriksson, L. 2008. Governance-by-standards in the field of stem cells: Managing uncertainty in the world of 'basic innovation'. *New Genetics and Society*, 27, 99-111.

PART 1
Changing Boundaries of Human, Nonhuman and Society

Part 1 will take an explicit look at the boundaries governing our contemporary thinking about life, or the boundaries between human and nonhuman and related modes of living.

Starting with one of the most prevalent boundary between humans and nonhumans – the question about the human and the animal – Tora Holmberg and Malin Ideland explore human-animal relations in the context of transgenic mice. They take a look into silences that create or maintain particular categorizations of living beings and modes of suffering. The relationship in question here is the question of ethics of care in relation to experimental animals in the context of new medical technologies. Focusing on mostly unexplored terrain, or on "discursive silences", the analysis yields interesting insights into the attribution of care, its relation to suffering and life in nonhuman animals. The productive silences analysed here can be fruitfully contrasted to insights provided by subsequent chapters focusing on the logics of care in the context of human patients. The question of suffering relates differently to different animals and the ways of handling dilemmas resulting from the application of the universal ideals of good life across species boundaries remind us about the underlying speciesism in our prevalent logics of care in the laboratories of biomedical innovation.

The following chapter on pluripotent stem cells takes us to the heart of biomedical innovation from another angle. Focusing on the relation of (political) definitions, material practices and potentialities invested in new bio-objects such as stem cells, Lena Eriksson points to tricky problems in managing new forms of life. Pluripotent stem cells are partly defined by politics surrounding experimental practices, partly by material practices that make possible the complex configurations needed to produce viable forms of life. Her analysis about the relations between political regulation and material practices reveals how problematic it is to draw boundaries between epistemology and ontology, or between politics and corporeally potent matter, as they co-configure each other in the case of pluripotent stem cells.

Ragna Zeiss' chapter on water takes the question on epistemology and life even further from here. Taking water as the focus of her chapter she asks why is it so hard to think of water as an object of life even if has been a matter of concern to life for organized societies. The problem of framing water as bio-object seems to stem from the fact that it is only one component of life and as such "too small" in terms of chemical make up of being life itself. At the same time, however, it is too big as an analytical category as always-already-there for life, exceeding it and being a substance that has its own "life" beyond organic living beings. As such, water demonstrates both limits of empirically attributing life to anything that lacks water at molecular levels and at the same time shows how the whole current epistemological paradigm of biology at large presuppose water before life (on earth and elsewhere) and does not contest the role of water in life. Therefore, water has not yet been subject to bio-objectification but could be so in the future.

Conor M.W. Douglas' chapter closes the first part of the book by questioning some of the leaky boundaries between human, technical and natural in the context of research subjects and processes by which they are rendered as objects in the experimental procedures of bio-medical science. In his analysis, research patients are not only known and represented in particular ways depending on different disciplines participating in clinical research, but also intervened as different objects of research with different apparatuses. In his case the patient becomes a set of genes, a site of biochemical reactions, an object of economics and so on depending which discipline takes the life of the patient as its object of intervention. Here, as in preceding chapters of the first part of the book, the question about "life" as a corporeal living subject and as a research object becomes intellectually approachable and materially tangible through pre-established categories of practice. These, as all of the chapters in the section attest, are becoming increasingly contested as the research moves onward in producing bio-medical facts/artefacts and creating new ways to represent, intervene and produce living beings.

Sakari Tamminen and Niki Vermeulen

Chapter 1

Challenging Bio-objectification: Adding Noise to Transgenic Silences

Tora Holmberg and Malin Ideland

Mapping the problem

There seems to be an endlessly rehearsed theme when it comes to transgenic and genetically modified animals: in research and in public discourse, they are portrayed as sources of future salvation from human illnesses. Transgenic animals, and thus the discourses surrounding them, embody hope and expectations of future, scientific breakthroughs. Because of its frequency, we would like to characterize this theme as transgenic *noise*. However, there are also striking *silences* when it comes to ethical and welfare concerns. We will address the noises and silences through the analytical lens of *bio-objectification*, and discuss how this particular bio-object – the transgenic mouse – can be viewed as both the product of bio-objectification, and as a potential challenger of the same process.

To give some background, transgenic animals are animals that have been genetically altered on purpose; genes have been either knocked-out, added or reinforced, in order to study the effects in a living (or dead) organism. Although the majority of animals being used are mice, many other species, including sheep, rats, fish and pigs, have recently been modified. In Sweden, which provides the national context of this chapter, about a third of all mice used in research are transgenic in one way or the other (Swedish Board of Agriculture 2008). Transgenic animals are produced and bred all over the world, both commercially and at special transgene units on university campuses, and are thus subjected to what can be labelled "technoscientific bespoking" (Michael 2001). By making animals "ready-to-order", the risk of instrumentalization is pressing. Through standardization, homogeneity is an important dimension of the bio-objectification process.

The structural conditions that frame the handling of transgenic mice include the legal context, the international, national and local policy regulation and the ethical review process. First of all, the Animal Welfare Act states that experimental animals should not be subjected to "unnecessary suffering" (Djurskyddslagen 1988: 534). Since 2007, Swedish experiments on animals are controlled by the Board of Agriculture. Local animal ethics committees scrutinize projects dealing with animal experiments through a mandatory reviewing process. Concerning transgenic animals, the application must state if it concerns "genetically modified animals", and declare if this means that certain welfare effects are expected.[1] Examples of problems specific to transgenic animals described in the literature concern the "surplus animals" that are used in production and breeding, the physical and emotional burden put on the animals used for production (especially on the female donors and surrogates), and the prevalence and risk of unexpected phenotypes (Schuppli, Fraser and McDonald 2004). Other difficulties specific to transgenic animals are assessments of the phenotypic and welfare status of genetically modified (GM) animals while the establishment of central databases with associated data to phenotypes is also problematic (Nuffield Research Council 2005).[2]

Furthermore, transgenic animals constitute forms of techno-scientific hybrids, and, as such, simultaneously challenge and confirm cultural categories and dichotomies (Brown et al. 2006). They can be understood as "boundary crawlers", as critters constantly balancing on the fine line between nature and culture, animal and human, organism and innovation, reality and model, science and technology. In other terms, these bio-objects have the potential of shifting a number of boundaries. Or, as Haraway so aptly puts it: "transgenic creatures, which carry genes from 'unrelated' organisms, simultaneously fit into well established taxonomic and evolutionary discourses and also blast widely understood senses of natural limit" (Haraway 1997: 56). They can be viewed as heterogeneous, culturally contested objects, and therefore highly interesting for any cultural science studies endeavour. However, one can always ask what is in it for the animals? What if the mice in our study could speak? What would they add to the discourse? In this paper, we will try to speak *nearby* mice (Hayward 2010), as an ethical standpoint, meaning that we will try to bring them into the text by way of asking the questions above throughout the chapter.

With this background in mind, we address the politics of bio-objects – the contingent positioning – by highlighting how dilemmas with transgenic animals become constructed as a non-issue by the people who handle them in practice; laboratory workers and members of animal ethics committees. We will do so in the

---

1   When the application has gone through the review process – which it usually does since 99 per cent of all applications get approved (but often with certain conditions [Nordgren and Röcklinsberg 2005]) – it works as a tool for researchers, animal technicians and veterinarians to decide on treatments and procedures.

2   The formal ethical apparatus apart, each and every one who works within the area (researchers at different levels, students, laboratory assistants, animal technicians, but also lay people in the animal ethics committees) have to take individual stands in order to handle daily situations and dilemmas: When is an animal too unhealthy, where is the endpoint? When and how is the mouse to be euthanized? What is a good enough aim of the project, justifying the experiments that I do? What is morally correct, and where do I draw the line? These questions have no given answers. They cannot easily be solved by a cost-benefit analysis, through referring to the Three Rs (refinement, reduction and replacement) or through ethical guidelines. Ethical dilemmas are constantly present and their character change depending on context, and in relation to the animal in question (Haraway 2008: 71).

empirical part, by recapturing some of the lessons that have been learned from our project, showing how narrations of the transgenic animals as any other animal are – at the same time – exclusive and different, and create a space for both transgenic silences and noise. A tension between the normal and the different is built up and adds to the discourse on transgenic mice as ordinary, but at the same time quite extraordinary – as *ordinary treasures* (Holmberg and Ideland 2009). In the last section, we take the analysis a step further to explore how the bio-objectification process can be interrupted and challenged, and invite the "trans" of transgenics to do some analytical work, as we consider these silences in the light of conversations with theoretical encounters with "trans" in the writings of feminist science studies scholars.

Some notes on method

As already mentioned, the ambiguous character of transgenic animals, captured so well in Haraway's quote above, rarely becomes articulated, neither in laboratories and ethics committees, nor in interviews with people who represent these arenas. Perhaps the most striking result of the research on which this chapter is based is that people who work with and/or ethically review research with transgenic mice, seldom articulate that there are any specific ethical issues for this branch. Transgenic animals have thus not become an "issue", in contrast to for example genetically modified crops, cloned animals or genetic tests. This silence can be viewed as a discursive effect of institutionally produced exclusion procedures (Foucault 1972), rendering some matters "unspeakable" (Billig 1999, Kulick 2005). The question is how these discursive silences take shape and what rhetorical devices and strategies contribute to transgenic silences. In order to answer these questions, we proceed from the discursive perspective that focuses on how people make use of different interpretative repertoires or culturally available sets of statements. Wetherell and Potter (1987, 1993) argue that people make use of specific interpretative repertoires and construct versions of reality in relation to the social context. Speech is viewed as a social act, rhetorically organized to bring to the fore a contextually fitting version of reality (Potter 1996). Included in the social situation are, besides participating (human and non-human) actors, assumptions about what the "correct" opinion might be, and overall cultural norms and values. Brown and Michael (2001), for example, show how scientists and medical practitioners routinely switch between different repertoires in their legitimizing of porcine organ donors for humans. Depending on the comparison point – human or non-human primates – different cultural-ethical or natural-scientific repertoires are used, all with the effect of proving that from a scientific point of view pigs are similar enough to humans to be used as organ donors, but morally speaking different enough to be used as a living organ bank.

The project this chapter draws on consists of two case studies, with empirical data collected through ethnography, including observations and interviews (see

Marcus 1995). The first specific case henceforth referred to as Laboratory Workers (LW), conducted by sociologist Tora Holmberg, explores the research practice and considers how researchers, laboratory assistants and animal technicians handle dilemmas in talk and practice.[3] In the second study, ethnologist Malin Ideland investigates how members of animal ethics committees talk about transgenic dilemmas, both in committee meetings and in individual interviews (Ethics Committees Members, ECM).[4] The empirical data from the research project, as a whole, consists of 40 semi-structured interviews and a large number of observation protocols from ethnographic field studies. Both interviews and observations have focused on how researchers and members of the committees handle dilemmas with animal experiments in general and transgenic animals in particular. From this data we have constructed categories of rhetorical strategies, commonly used in the empirical material.

Trangenic silences – ordinary mice and normal transgenes

It was mentioned in the introduction that transgenic mice can be mail-ordered from commercial as well as university based breeders. Phenotypic consequences can be rather tricky to predict when the animal model has been locally produced, and different tests to characterize early behavioural and other deviations have consequently been developed (Nuffield Council of Bioethics 2005). Most of the interviewees agree that there seldom occur any dramatic and unexpected effects, which in itself can become a problem: "You have the dilemma I have seen that the mice I have are so immensely normal, I look carefully for the smallest deviance and I don't find a thing" (Interview researcher, LW).

What is good for the animal – in this case a lack of unexpected effects – is not always good for the experiment, which puts the researcher in a dilemmatic situation. On the one hand you aim at certain effects, and on the other you might have to break off the experiment if they occur. While uncertainty concerning phenotypic consequences is a problem some of the interviewed ethics committee

---

3   Data is diversified and consists of observations from a two-week course for researchers (Holmberg 2008), field work with several research groups, including observations of animal experiments in practice, and field work at two different animal houses – departments where the lab animals are bred and kept. Moreover, in total 20 semi-structured interviews with researchers at different levels, laboratory assistants and animal technicians from two different Swedish universities have been conducted (Holmberg 2010).

4   Ideland interviews 20 members of animal ethics committees during 2006 and 2007. Among these 20 members, three persons represent animal welfare or animal rights organizations, six are representatives of political parties, ten are scientific experts from different disciplines and one is an animal technician; all in all, six different local animal ethics committees are represented. The interviews were semi-structured. Twelve meetings, both in preparatory and plenary meetings, in six different committees have been observed (Ideland 2009).

members commented on, more often they emphasize the transgenic *normality:* "they act and look like any other mice" (Interview animal technichian, ECM). Comparisons with "ordinary mice" here mean that the transgenic animal becomes normalized – it is nothing but yet another mutated mouse: "If you have two cages, one with wild types or ordinary animals, and one with transgenic white mice – you cannot tell the difference. [...] They eat normally, they live normally, and they breed normally" (Interview researcher, ECM).

These mice pass as ordinary and their normality becomes established primarily by means of visual representations and welfare assessments work mainly on this level. Comparisons with "ordinary mice" also mean that transgenic animals are somewhat naturalized. It can seem paradoxical to describe transgenic mice – a symbol of high-technological innovations – as essentially natural. Interviewees portray the modification of the genome as one end of a continuum and as such, as nothing new. "Nature" has always been "doing" similar changes through spontaneous mutations, and nowadays it is done in the laboratories, in an accelerated process: "The difference from spontaneous mutations is that now we are speeding up the process. That's really the only difference" (Interview researcher, LW). The prevalence of spontaneous mutations is often brought up as arguments in the defence of genetic modification. The transgene technique is legitimized since natural processes can entail similar results as conscious technological interferences into biology. Discussions about naturalness plays down and normalize gene technology (see Butler 1990). It is also obvious that referring to nature "hides" human agency, it just happens. Moreover, it hides animal agency too, along with the labour these critters do. The "bio-part" of the bio-object is here emphasized in terms of uncontrollable nature. Somewhat paradoxically, this use of nature and naturalness is at the same time objectifying the mice, since it legitimizes the genetic modifications.

A related rhetorical instrument is history, narrated in the study as claims that humans "always" have been affecting these kinds of changes, not least through breeding:

> We humans have during thousands of years domesticated animals and changed their behaviour through selective breeding. An artificial selection rather than a natural one. So I am not outraged by the manipulation, the meaning of genetic modification in laboratories. Many times the effect isn't more than through breeding. It is a little faster. [...] This extra thing we do in the lab, I don't consider it ethically difficult, that we play God and invade the animals. (Interview researcher, ECM)

Reference to breeding is a common rhetorical technique legitimizing genetic modifications, when it comes to both animals and to plants. Once again this strategy contributes to placing the technology in a historical context. By selectively using a history of science it is possible to avoid the fear and insecurity that often follow new technologies (Marwin 1988). But history can also be used in the opposite way; to create fear of new techniques. Gene technology has, for example, been connected to eugenics and Nazi experiments. History can be used to either criticize or legitimize a technique. In this case, the history of transgenic animals is mostly used to point to the normality and naturalness of these mice.

An apparently opposing rhetorical strategy is to alienate a new technology from one that is already established. The normalization process of the transgenic animals – the construction of them as ordinary – also means that transgenic mice are described as more ordinary than other – more or less debated – phenomena. Several researchers oppose the ongoing discussion of transgenic animals' possible welfare problems, through comparisons with other activities:

> If you change the genetic make up so that … the phenotype of the animal is not feeling well, or doesn't survive in the long run or so, then of course it is unpleasant for that animal. It has to be. But that does not only concern experimental animals, you have the debate about Belgian blue, for example, these large meat cattle which are not really genetically modified but bred that way. (Interview researcher, LW)

In defending the use of animals in research, it is customary to talk about how it actually is worse in other sectors, not least in the meat industry (Birke, Arluke and Michael 2007). In several interviews the cattle Belgian Blue is used as a deterrent example of problematic phenotypes caused by "common" breeding. It is possible to understand this contrasting as a discursive strategy to handle inherent dilemmas with the use of animals in research; like the researcher quoted above who does *not want* the mouse to be affected by unpleasant effects, but still *wants* to be able to study these effects as a result of his research. It also illuminates two parallel processes of bio-objectification. Through the process of homogeneity the mice become standardized instruments for research, and at the same time, the heterogeneous process constructs them as living organisms – homogeneous due to human interventions and dissimilar at the same time. As such they introduce mess in the otherwise "tidy" experimental practice and discourse (Holmberg 2010). They are synchronously standardized organic models – products of trans-procedures – and living, sensing mice – moral categories – and individuals working with this product/life are more or less forced to handle the ambivalence through discursive as well as other practices.

It seems clear that comparisons with ordinary, inbred laboratory mice, contribute to the construction of "normal", non-deviating transgenics, in which for example problems of unexpected phenotypes are minimized. By way of naturalizing the transgene production, any welfare consequences are ultimately placed on the individual animal, not as a result of the transgenic technologies. Comparisons with other animals and businesses, showed how the shift of repertoires enables the portraying of the "good life" of transgenic mice and minimising the risks of genetically modified laboratory animals. In comparisons with other lab mice, as well as with other animals and fields of GM applications, suffering again appears as a problem with the mouse, not as caused by technologies. Human as well as animal agency moves out of sight, by way of the repertoires used. This hiding of agency is part of the "blackboxing-process", in which the process that lies behind scientific results becomes invisible (Latour 1999). The social factor should not be decisive for valid scientific data; matters of fact are regarded as a "mirror of nature" (Shapin and Schaffer 1989; see also Kruse 2006). In the same way successful – or unsuccessful – transgenic animal models are not represented as results of human agency. Instead, they are constructed as results of scientific progress, and in the black-boxing process human involvement, and thus responsibility, vanishes.

Transgenic noise – animals as treasures

As stated above, in comparisons with other animals and crops it is the normality of transgenic mice that is emphasized. However, when the interviewees instead compare the animals with humans, in particular, patients, another perspective appears. The transgenes then become transformed, through a shift in repertoires, from ordinary, common and safe, to cherished treasures. Thus a discursive shift takes place, but the meaning remains; there are no specific dilemmas with transgenic mice – they are like other laboratory mice – except the hope that they can "perhaps" open doors for future breakthroughs. They are ordinary, but still treasures.

> I: If we can create a drug which uses this gene product, then we can help all people with this disease, it could be Parkinson's disease or Alzheimer's disease, so this kind of knowledge is extremely critical. [...] So, because of this it is so ... precisely this with genetics and animal models are so awfully important. And it was rewarded with the Nobel Prize. (Interview researcher, LW)

The future is central in the talk about transgenics. Just like the deployment of history, the future is a major element in the legitimizing process, as it has been in many other, similar debates about new biotechnologies (Borup et al. 2006, Brown and Webster 2004). Hope for medical solutions become embodied in the experimental animals, but also in comparisons with humans – through the talk about future patients. The researcher quoted above, emphasizes future possibilities by referring to earlier research. Another researcher brings forward the potential possibility of the "risky" basic research.

> Medical progress has often not happened when you have looked for it intentionally. It has been a side-effect of a more general knowledge-searching. [...] you don't know the outcome. Perhaps nothing. Perhaps cure for all diabetics for their entire lives. Millions of people can be benefitted from it. (Interview researcher, ECM)

In this uncertainty scenario, the utility aspect that is supposed to be included in the cost-benefit-analysis becomes extremely plastic. This plasticity results in the near impossibility of questioning the purpose of the experiment, since it may perhaps be useful in the future. "Perhaps" is a key sign, a reoccurring word in the discourse concerning transgenic animals, not least in comparisons with humans. The animals will "perhaps" develop a problematic phenotype. They can "perhaps" lead to a cure for millions of patients. The uncertainty is both good and bad, but it is the good prospects that most often win, not least because there are patients who embody the expectations. Michael (2000) points out that these expectations can be viewed as *performative*. Through different discursive forms such as metaphors, narratives and other signs of promises of scientific progress and better (healthier) lives, they conjure up the future in the present (Brown et al. 2006: 2). In discourses on transgenic animals, hope for medical breakthroughs are being built in. "We cannot close those doors", is an effective argument to overshadow fears of the unexpected and the potential suffering for the animals. The unexpected is instead to be expected (Borup et al. 2006: 295), it is what can open doors hitherto closed, to further understanding and, in the extension, methods for treatments. In the negotiations that go on in the ethics committees, any purpose concerning possible scientific breakthroughs are good enough to justify animal experiments (Ideland 2009). What about the transgenic mice? Would they agree that this uncertainty scenario is a good enough reason to invest their lives and efforts – that is, their transgenic labour? Their voices become silent in the transgenic noise surrounding (possible) scientific success and progress.

Comparisons with humans, in particular patients, contribute to an economy of hope and scientific breakthroughs – where science as salvation curtails the suffering of animals – thus minimizing potential and actual ethical dilemmas. In the public discourse, biotechnologies are almost synonymous with the language and imagery of futuristic breakthroughs. In this discourse medical rationality is always a good enough reason to use animals, as well as for example embryos, for health purposes (Brown 2003). The hope for biotechnological salvation leads to an instrumentalization of animals. The animals lose their own value, and become dispensers for hopes for medical solutions, for "the good life" (Michael 2000, Brown et al. 2006). Now, instrumentalization, says Haraway, is intrinsic to embodied laboratory work, and it probably can not be in any other way (Haraway 2008: 71). Ambiguity is, as stated in the Introduction, inherent to the use and understanding of these mice, and refers back to transgenic animals understood as boundary crawlers, here balancing on the line between subject and object, individual and model.

Troubling trans

Taken together, what we have done so far is to investigate how competing discourses – in comparison to other phenomena – have been shaped and reshaped in conversations about transgenic animals. The message – that the transgenes are at once ordinary entities and radical exclusivities – is the same throughout, but different rhetorical comparisons are made that legitimize the transgenic enterprise. With the help of comparisons with ordinary laboratory mice, other animals and GM fields of applications, natural processes and humans, the increased use of genetically modified animals appear as all together unproblematic. It is rather constructed as an obvious continuation of both natural and historical processes, leading toward better treatments for future ill people, making the production and handling of experimental animals appear as totally agreeable. By way of promoting the potential medical benefits for humans, interviewees' statements add to a discourse on biotechnological salvation (Haraway 1997, Brown et al. 2006). The flexible use of – and frequent shifts between – interpretive repertoires, highlight cultural norms concerning ecology, human health and animal suffering in an interesting mix. It also highlights how this particular bio-object is a) actively constructed, and b) contextually contingent. Thus, the bio-medical discourse bio-objectifies the transgenic mice as either models or objects of salvation, or as living organisms. We argue that one of the features of bio-objectification processes in general is precisely the highlighting of certain dimensions, and silencing of other aspects.

A more important point, however, is that these flexible characteristics are not innocent, the shifts between different repertoires contribute to the construction of "transgenic silences". For example, there are suffering and thus welfare issues connected to the transgenic technology, like expected and unexpected phenotypes, and surplus animals. "Trans" can thus be viewed as a special kind of suffering. But this specialness is hidden under a rhetoric of commonness, which contributes to how actors manage cultural and ethical dilemmas. For a democratic, responsible public science debate, which can intervene and raise new concerns, something completely different from silence is needed. In the present case, we will re-instate "trans" as a troubling tool, with the aim of counteracting transgenic silences. In order to do so, we will dwell on the concept of trans. Can "trans" as a pre-fix provide some hope against GM hype?

As a standard definition, "trans" derives from Latin and connotes to "across or over; beyond or above; from one place to another; to cross over, pass through, overcome" (Pryse 2000: 105). It transcends certain limits, and is thereby "suggesting the unclassifiable" (Hayward 2008: 253). Haraway (1997) writes that "trans" is both the process and the product of the crossing over of nature/culture borders. Trans "cross a culturally salient line between nature and artifice, and they greatly increase the density of all kinds of other traffic on the bridge between what counts as nature and culture" (1997: 56). Haraway's conception of the transgenic and other mice as next of kin (Haraway 1997) enables us to bring some transgenic trouble out of the closet. As already mentioned several times, transgenic animals are boundary crossing in many ways, and cannot easily be categorized. They are therefore related to other queer and culturally disturbing figures like for example the "female man" and "feral children". Haraway states that a transgenic mouse is many things, and inhabits numerous cultural spaces and meanings. For example, it is a model for a disease, a living animal, a commodity, a machine, an organism, a tool, a patented animal and an invention, bio-objectified in various ways and through various discursive means. In that sense, the transgenic rodent shares the multitudes of meanings with the ordinary inbred mouse (see, for example, Birke 2003). They are both messy. But the mouse is also, as noted from one of our case studies, involved in laboratory interaction and as such an actor, not only responding to the experimentalist's actions, but also – although in a very limited way – actually shaping his or her behaviours and feelings (Holmberg 2008). The trans-thing that Haraway develops in her study, threatens to lead away from what actually happens in interaction, to a more cultural level. And what happens to cultural norms when the frequent use of transgenic animals change their status from the exception that they were in the early 1990s, to being part of the norm, as they are today?

"Trans" in Franklin's version is inscribed in the term *transbiology* – describing the contemporary organization or rather reorganization of living matter, of what Foucault named "life itself" (Franklin 2007). Transbiology is not just an epithet for laboratory action, but also captures the postmodern diffusion of science into all imaginable spheres of society; popular culture, politics, economics etcetera. Franklin builds on the trans-concept of Haraway, and suggests that in the same way as the cyborg was helpful to understand the contemporary couplings of biology, technology and informatics, transbiology can be used as a figurative trope, a tool to understand today's norm in biology – as something "not only born and bred, or born and made, but made and born" (Franklin 2006: 171). Transbiological offspring – such as Dolly, the cloned sheep, or indeed transgenic mice – were at first miraculous because they were so normal. What made Dolly a successful clone, paradoxical as it may seem, is that she was both common and unique. In a similar vein, one could argue that there is something special with the transgenic mice – but that it is hidden under a rhetoric of commonness and "business as usual", which contributes to the management of inherent dilemmas. But if we take a step further and consider the content of the comparisons made, a clear point appears: Are the transgenic mice "really" so different? Are they not just, as our interviewees state over and over again, the contemporary endpoint of a beaten track of domestication of other species? The consistent and rehearsed argument has been that transgenic animals are understood as both ordinary, domesticated laboratory mice, and as all together exclusive, the ultimate sign of biotechnological progress. Transgenic mice thus inhabit a flexible meaning, a doubleness that fits nicely with a discourse of biological control. As Franklin writes:

> In this view, the genetically engineered animal is both a symptom of human overconfidence in biological control and the culmination of a lengthy process by which the drastic consequences of domestication has been unfolding. (Franklin 2007: 31)

Halberstam has used Franklin's transbiology concept in a fruitful way, investigating knowledge production taking place outside of the laboratory; in wild life films and animal animations, as in horror movies (Halberstam 2008). She states that the concept highlights the transgressive intervention going on, in which traditional views of sexuality, genealogy, body and reproduction become challenged. In a similar way of reasoning as Franklin and Halberstam, and indeed Haraway before them, one could argue that the transgenic mice do challenge many given norms concerning all of the above, but also in regards to kinship and species boundaries. Nevertheless, in our view, trans in feminist science studies tends to be understood as something inherently positive. But what is so good about it for the mice, or rats, or sheep? And, as some critics argue trans can as easily be interpreted as a nightmarish manifestation of masculine ideals of transcendence and limitlessness (Birke 1999: 164). Trans is not something essentially good – and Haraway is clear on this point – trans also needs ethical considerations.

Braidotti writes in her book *Transpositions* (2008) – a concept borrowed from Evelyn Fox Keller's reading of Nobel prize winner Barbara McClintock – about "trans" as admitting "alternative ways of knowing" (Braidotti 2008: 6), that is, both epistemological and ethical issues are at stake. Braidotti advocates a post-humanist, nomadic perspective in which transpositions stands for a sustainable ethics (2008: 33). To be nomadic, that is in transition, is also to be somewhere, it does place the actor outside of history. Transpositions, in Braidotti's words, are about becoming rather than of being.

Thus, we read Braidotti here as saying that one should not worry too much about technology altering nature but focus on the relationship as constitutive of species and individuals. Now, what would a Braidotti perspective bring to the problem of transgenic silences? Sustainable ethics, in our reading, is partly about asking new questions about transgenic animals; not only about whether they are normal or natural, how one should assess deviant behaviour or if they can escape from the lab and pollute wild rodents (these questions are of course also important), but also about what kinds of technologies (both technical, rhetorical and institutional) that enables these animals, how they affect the understandings of what it means to be human and mouse, and of the practices involved in producing these understandings. What we would like to see more of are the mice; their suffering, their agency, their labour. Haraway concludes that the Oncomouse – a transgenic mouse carrying e.g. breast cancer genes – is an incarnation of the Christian Messiah (Haraway 1997). She bears our suffering so that we (perhaps!) can be healthy. Within the sacrifice logic that is foundational in animal experimentation (Lynch 1988, Birke, Arluke and Michael 2007), the transgenic mouse carries the suffering of future patients. But, does this suffering need to be silent? Can she not scream and shout? Do the experiments need to be hidden and performed in secrecy? (Holmberg and Ideland 2010) We would argue that the labour and the suffering that is put into the transgenesis by the mice, should be part of the noise rather than the silence.

To conclude, our analysis shows 1) how bio-objects become discursively constructed with the help of rhetorical strategies, by which some aspects are highlighted while others are silenced, 2) how processes of homogeneity and heterogeneity work simultaneously, framing the transgenic mouse differently, and 3) how the bio-objectification process, through a third dimension of transgeneity, can be challenged by the very objects it produces. In other words, we add to the understanding of the bio-objects and their ability to change the process from within. Transgeneity in our case, means the process through which the ambiguous identity of the bio-object work to produce agency and new subject positions. Moreover, we would like to see a little more of trans in the public and scientific debates, since it might counteract transgenic silences. Bringing the "trans-thing" back in, is thus a way of highlighting, rather than hiding, dilemmas with transgenic animals. It may also be a way of bringing the mice into the discourse. As Eva Hayward notes, animals are always "troubling the language that attempts to name them. In this way, non-human animals seem to put an oral void into language. Animals cannot be named without invoking the limits of the process of naming" (Hayward 2008: 260). Thus, trying to speak nearby transgenic animals means reinstating "trans", not as an attempt to specify a new category – trans is still about the unclassifiable – but to invite mice to crawl into discourse and create some noise.

Acknowledgements

The authors would like to thank Måns Andersson, Conor M.W. Douglas, Eva Hayward and Helena Wahlström for valuable and thorough comments on an earlier version of this paper. We are also grateful for constructive reviews from the editors of this book.

References

Billig, M. 1999. *Freudian Repression: Conversation Creating the Unconscious.* Cambridge: Cambridge University Press.

Birke, L. 1999. *Feminism and the Biological Body.* Edinburgh: Edinburgh University Press.

Birke, L. 2003. Who – or what – are the rats (and mice) in the laboratory? *Society and Animals,* 11(3), 207-24.

Birke, L., Arluke, A. and Michael, M. 2007. *The Sacrifice: How Scientific Experiments Transform Animals and People.* West Lafayette, WV: Purdue University Press.

Borup, M., Brown, N., Konrad, K. and Van Lente, H. 2006. The sociology of expectations in science and technology. *Technology Analysis and Strategic Management,* 18(3-4), 285-98.

Braidotti, R. 2008. *Transpositions.* Cambridge: Polity Press.

Brown, N. 2003. Hope Against Hype – Accountability in biopasts, presents and futures. *Science Studies,* 16(2), 3-21.

Brown, N. and Michael, M. 2001. Switching between science and culture in transpecies transplantation. *Science, Technology, & Human Values*, 26(1), 3-22.

Brown, N. and Webster, A. 2004. *New Medical Technologies and Society: Reordering Life.* Cambridge: Polity Press.

Brown, N., Faulkner, A., Kent, J. and Michael, M. 2006. Regulating hybrids: "Making a mess" and "cleaning up" in tissue engineering and transpecies transplantation. *Social Theory & Health*, 4(1), 1-24.

Djurskyddslagen 1988. 534 (Animal Welfare Act) [online]. Available at: http://www.notisum.se/rnp/SLS/lag/19880534.htm [accessed 16 June 2010].

Forsman, B. 1993. *Research Ethics in Practice: The Animal Ethics Committees in Sweden 1979-1989.* Göteborg: Centre for Research Ethics.

Foucault, M. 1972. *The Archaeology of Knowledge & The Discourse on Language.* New York: Pantheon Books.

Franklin, S. 2006. The cyborg embryo: Our path to transbiology. *Theory, Culture and Society*, 23(7-8), 167-88.

Franklin, S. 2007. *Dolly Mixtures: The Remaking of Genealogy.* Durham, NC: Duke University Press.

Halberstam, J. 2008. The nanoengineering of desire, in *Queering the Non-human*, edited by N. Giffney and M. Hird. Aldershot: Ashgate Publishing.

Haraway, D.J. 1997. *Modest_Witness@Second_Millenium. FemaleMan©_Meets_ OncoMouse™.* London and New York: Routledge.

Haraway, D.J. 2008. *When Species Meet.* Minneapolis: University of Minnesota Press.

Hayward, E. 2008. Lessons from a star fish, in *Queering the Non-human*, edited by N. Giffney and M. Hird. Aldershot: Ashgate.

Hayward, E. 2010. FingeryEyes: Impressions of cup corals. *Cultural Anthropology*, 25(4), 577-99.

Holmberg, T. 2008. A feeling for the animal: On becoming an experimentalist. *Society & Animals*, 16(4), 316-35.

Holmberg, T. 2010. Tail Tales: How researchers handle transgenic dilemmas. *New Genetics and Society*, 29(1), 37-54.

Holmberg, T. and Ideland, M. 2009. Transgenic silences: The rhetoric of comparisons and the construction of transgenic mice as "ordinary treasures", *Biosocieties*, 4(2), 165-81.

Holmberg, T. and Ideland, M. 2010. Secrets and lies: "Selective Openness" in the apparatus of animal experimentation. *Public Understanding of Science*, published online, 26 August.

Ideland, M. 2009. Different views on ethics. How animal ethics is situated in a committee culture. *Journal of Medical Ethics*, 35(4), 258-61.

Jordbruksverket (Swedish Board of Agriculture) 2008. *Användningen av försöksdjur under 2007* (Use of experimental animals 2007), Report, Dnr: 31-808/08. Stockholm: Jordbruksverket.

Kruse, C. 2006. *The Making of Valid Data: People and Machines in Genetic Research Practice.* Linköping: Department of Technology and Social Change, Linköping University.

Kulick, D. 2005. The importance of what gets left out. *Discourse Studies*, 7, 615-24.

Latour, B. 1999. *Pandora's Hope: Essays on the Reality of Science Studies.* Cambridge, MA: Harvard University Press.

Lynch, M. 1988. Sacrifice and the transformation of the animal body into a scientific object: Laboratory culture and ritual practice in the neurosciences. *Social Studies of Science*, 18(2), 265-89.

Marcus, G. 1995. Ethnography in/of the world system. The emergence of multi-sited ethnography. *Annual Reviews of Anthropology*, 24, 95-117.

Marwin, C. 1988. *When Old Technologies were New: Thinking about Electric Communication in the Late Nineteenth Century.* Oxford: Oxford University Press.

Michael, M. 2000. Futures of the Present: From Performativity to Prehension, in *Contested Futures: A Sociology of Prospective Techno-Science*, edited by N. Brown, B. Rappert and A. Webster. Aldershot: Ashgate.

Michael, M. 2001. Technoscientific bespoking: Animals, publics and the new genetics, *New Genetics and Society*, 20(3), 205-24.

Nordgren, A. and Röcklinsberg, H. 2005. Genetically modified animals in research: An analysis of applications submitted to ethics committees on animal experimentation in Sweden. *Animal Welfare*, 14, 239-48.

Nuffield Council of Bioethics 2005. *The Ethics of Research Involving Animals.* London: Nuffield Council of Bioethics.

Potter, J. 1996. *Representing Reality: Discourse, Rhetoric and Social Construction.* London: Sage.

Pryse, M. 2000. Trans/Feminist Methodology: Bridges to Interdisciplinary Thinking. *NWSA Journal*, 12, 105-18.

Schuppli, C., Fraser, D. and McDonald, M. 2004. Expanding the Three Rs to Meet New Challenges in Humane Animal Experimentation. *ATLA*, 32, 525-32.

Shapin, S. and Schaffer, S. 1989. *Leviathan and the Air-Pump – Hobbes, Boyle and the Experimental Life.* Princeton, NJ: Princeton University Press.

Wetherell, M. and Potter, J. 1987. *Discourse and Social Psychology.* London: Sage.

Wetherell, M. and Potter, J. 1993. *Mapping the Language of Racism.* New York: Columbia University Press.

Chapter 2

Pluripotent Promises: Configurations of a Bio-object

Lena Eriksson

In this chapter, I examine a bio-object that has pushed and prodded the boundaries for our understanding of both the conception and continuation of life – the human embryonic stem cell (hESC). If bio-objects are forms of becoming, the hESC is a bio-object par excellence in that it is the very definition of becoming. In fact, it is this definition of becoming that is the investigative focal point of this text. A central trait in the hESC configuration is the cells' inherent *pluripotency*, and this trait is analytically disentangled and traced as the hESC bio-object emerges and circulates through different contexts.

So why a bio-object framework, rather than something else? The hESC could alternatively, for example, be understood as an example of what Mol (2003) would call 'ontic practice' in an unusually material sense. But while Mol's analysis turns away from epistemics in favour of the multiplicity of practices that makes and breaks the object of actors' attention (in Mol's case Atherosclerosis), the notion of the bio-object presents us with a material entity that is simultaneously epistemic and ontic practice. The bio-object also adds a valuable dimension to the 'onto-technical standards' examined in previous publications, in that it accentuates the 'thingness' of such epistemic practice (Eriksson and Webster 2008, Webster and Eriksson 2008). Furthermore it explicitly adds temporality to this epistemic materiality, thus making a bio-object epistemics, ontics, temporality and practice in a material wrapping.

Pluripotency, far from being a static and axiomatic trait in hESCs, is as adaptable and fascinating as the stem cells themselves. Its careful configuration has been central to the hESC's continuing relevance.

The concept of pluripotency tends to be taken for granted in stem cell research of all denominations, in social as well as natural sciences. Stem cells are interesting to science and medicine because they are not yet fully formed cells that still hold the potential to become a whole host of different cell types, thus making them potentially useful in several ways – as models of early development and disease, as material for toxicity screening and most famously as building material for regenerative medicine therapies. Embryonic stem cells are extra potent – they have a very wide developmental capacity. This separates them from adult, or somatic, stem cells that are already locked in to particular developmental routes. Pluripotency is commonly defined as the ability to differentiate into (almost) any human cell type in the body and embryonic stem cells could therefore, so says the theory, revolutionise regenerative medicine. Promise is thus the very stuff that these cells are made of – the pluripotent human embryonic stem cell is by definition a tentative future projection.

Briefly, pluripotency can be seen to work in several different ways. Pluripotency has served to protect a somewhat embattled research community sandwiched between two powerful discursive continents. Ethical objections from anti-abortion groups as well as queries regarding viability of hESCs raised by proponents of adult stem cell research can be countered by way of pluripotency. Pluripotency can also serve as a gold standard benchmark to uphold quality assurance and credibility within already accredited institutions, such as the UK Stem Cell Bank. Simultaneously there is concern that too ambitious a benchmark will exclude certain actors in the field, and may therefore be seen to be sub-optimal. Pluripotency-as-quality-assurance presupposes access to particular resources and skills, and by extension it outlines a particular laboratory infrastructure. Attempts to standardise practices and collectively characterise hESC lines has also accentuated the cells adaptability and raised the prospect of inadvertently materially creating a skewed hESC signature based on an idealised idea about what such cells should look like. Finally, pluripotency can, if configured to fit a more functionalist definition, usher hESCs into a clinical future (Eriksson and Webster 2008).

Chapter outline

To understand the varying functions of pluripotency, one must consider the different contexts of embryonic stem cells. The chapter starts with a brief description and analysis of embryonic 'stem cell history', making the point that the context in which embryonic stem cells were first conceived of changed radically over a short span of time. When the embryonic cells originated from mice only, their inherent developmental capacity remained ethically unproblematic and experimental practices followed a proof of principle logic. The more potency the cells could be shown to possess – mainly by way of conjecture – the better. Only when embryonic stem cells were derived from humans did the notion of potential come into practical and ethical focus and the pluripotency of the cells began to be carefully polished and policed. Pluripotent hESCs are 'just right' in terms of potential – they are not so potent as to make a baby but yet potent enough to be far superior to adult stem cells. This configuration of pluripotency – meticulously positioned between 'totipotency' and 'multipotency' – has been important to support the argument that hESCs simultaneously offer and restrict maximum potential promise.

When translated from the public sphere to the confines of individual laboratories, however, pluripotency as defined in its discursive public dress becomes a practical headache – it is very hard both to maintain and to ascertain. The 'public' stem cell is a discursive object often encountered both in popular media and scientific journals – it is depicted as a neat, undifferentiated cell that can and will readily differentiate into almost any cell type. The public stem cell is considerably more agreeable than the 'private' one; the latter can mainly be found in its material form in laboratories or as a discursive entity in closed scientific meetings. The private stem cell is a moody creature that will keep its handlers on their toes; it is difficult to maintain it in an undifferentiated state, let alone to push it down specific developmental trajectories.

The first configuration of pluripotency that positions the cells as 'just right' in terms of potentiality is a configuration of the public stem cell. Standardisation work however, deals with the private stem cell. This raises at least two problems: Firstly, if the 'bar' for proving pluripotency is set too high, many laboratories (and countries) could opt to go their own way and this would represent a serious set-back for those who wish to see an international scientific consensus in the field of human embryonic stem cell research. Secondly, the cells' tendency to adapt to surrounding material and practices raise some concern that too strict definitions of how a hESC should be defined and grown would not only risk excluding certain cell lines but to materially transform them. Thirdly, as the prospect of using the cells for therapeutic purposes is considered increasingly realistic, the 'public' configuration of pluripotency changes from being a discursive resource to instead become a clinical risk.

A brief and incomplete history of embryonic stem cells

To understand how pluripotency came to be such a central and carefully configured feature of the human embryonic stem cell, one needs to appreciate the disjuncture between what we might call today's historical stem cell and the historical stem cell of yesteryear. In November 1998, the journal *Science* published a paper in which a team of scientists purported to have derived the first stem cell lines of human origin (Thomson et al. 1998). Today's stem cell narrative tends to take this isolation of the first hESCs as its point of departure, referring to previous work on mouse embryonic stem cells as the linear and unproblematic past that enabled the present. This version of stem cell history renders invisible the discrete nature of mouse embryonic stem cell work as research with its own objectives, not necessarily aiming towards a future of human regenerative medicine.

The regenerative medicine platform is what has caught both media attention and imagination over the last decade and is arguably what comes to mind when stem cell research is mentioned today; the use of stem cells primarily as a way of repairing or replacing damaged tissue. This relatively recent understanding of stem cells has had a tendency to eclipse and obscure the longer history of stem cells as a model for human genetics. While stem cells have only resided in the media spotlight for a decade, they have a longer history as a tool to study targeted pathologies in vitro by way of so-called knock-out mice. This 'disease in a dish' model pervades all fields of biomedicine and knock-out mice are today a standard tool that does not necessarily lead our thoughts to stem cell research.[1]

Of mice and men

The shift from a mouse platform with its inherent bold proof-of-principle logic over to a human platform also entailed a shift from a relatively ethics-free zone, to ground saturated with religious, ethical and philosophical conundrums pertaining to the very core of humanness and personhood.

While much of today's human stem cell research builds on the more comprehensive understanding of the mouse model, the mouse research was not carried out with the aim to eventually move over to a human model. The mouse embryonic stem cell was however the template model that scientists used to extrapolate conjectures regarding hESCs and when speculating in regard to the cells' developmental capacity.

While scientifically considered a great achievement, the derivation of mice from embryonic stem cells had been comparatively uncontroversial. At the time, it was in scientists' interest to demonstrate the many and marvellous things these embryonic stem cells could do. Therefore, growing an entire organism – a mouse – from a single cell was a desirable scientific goal. As a heuristic template model for hESCs, however, such an experiment would later prove uncomfortable in raising questions regarding the status of embryonic cells.

The political potency of human stem cells – toti, pluri and multi

The most basic definition of a human embryonic stem cell is that it is an undifferentiated cell with two vital properties; the ability to self-renew and the potential to differentiate into almost any cell type in the body (NIH 2011). The hESC is therefore defined by its possible future(s) and it has over the years been a conceptual balancing act on part of scientists to ensure that such a future is full of promise, yet not cause for ethical concern. The NIH web resource on stem cells states that a *totipotent* cell is one that '… can give rise to all the cell types that make up the body plus all of the cell types that make up the extraembryonic tissues such as the placenta', whereas *pluripotency* is defined as the '… ability of a single stem cell to give rise to all of the various cell types that make up the body. Pluripotent cells cannot make so-called "extra-embryonic" tissues such as the amnion, chorion, and other components of the placenta'. *Multipotency*, finally, is the 'ability of a single stem cell to develop into more than one cell type of the body' (NIH 2006).

---

1   For a fascinating study of how the mouse became a standardised laboratory tool, see Rader 2004.

These definitions are strongly interdependent – pluripotent cells are not totipotent, but yet considerably broader in developmental scope than multipotent (adult) cells. In the early days of hESCs, however, it was assumed and acknowledged by scientists – including the two pioneering team leaders Thomson and Gearhart – that the cells were likely to be totipotent (Wade 1998). However it was felt that the ethical reasons for not performing a totipotency test were insurmountable. Such a test, when performed in mice, involves taking the inner cell mass from one blastocyst and injecting it into another in order to establish whether the resulting embryo would be a chimera, that is carry both lineages. Therefore, the question of whether *human* embryonic stem cell lines could be said to be totipotent would have to remain a theoretical one.

In the context of the 1998 American political landscape however, such a question could simply not remain unanswered, or 'unknowable' as a *New York Times* article called it (Wade 1998). At stake was the possibility of whether human embryonic stem cell research could escape an existing and for the field potentially crippling ban on federal funding for human embryo research. The question was, in a manner of speaking, how embryo-like were the cells? Could they be said to be equivalent to an embryo?

Not a baby – keeping the anti-abortion lobby at bay

The distinction between toti- and pluripotency became important under a particular set of conditions; in the US it is illegal to use federal funds for research 'in which human embryos are created, destroyed, discarded, or knowingly be subjected to risk of injury or death greater than allowed for research on fetuses in utero (…)' (Dickey-Wicker, 1996). When the 1998 announcement regarding the successful derivation of hESCs was made, the question soon arose of whether the cell lines would fall under this legislation. In December of that year, the then NIH director Harold Varmus appeared before a Senate subcommittee to make a statement regarding the recently isolated human embryonic stem cell lines, which he called 'the first human *pluripotent* stem cell lines' (Varmus 1999, author's emphasis).

By referring to the cells as pluripotent rather than embryonic, the NIH director indicated the scientific and semantic reasoning that was to follow. Varmus explained to the Senate that while cells from an early embryo were considered totipotent, once a blastocyst had formed the cells that were to make up a subsequent embryo had already separated from those cells needed to support the embryo's continued development, namely the placenta and other 'extra-embryonic tissues'. The inner cell mass of such a blastocyst that the human embryonic stem cells were derived from could therefore not be said to have a 'totally potent' or embryonic status, as they could not develop into a fetus if put back into a uterus. Instead, they should be considered 'pluripotent' (Varmus 1999).

The pluripotent status of the hESC was what could potentially make human embryonic stem cell research permissible in federally funded laboratories, using federally funded tools.

Two months earlier, both scientists responsible for the newly derived human lines had speculated in public that the cells were totipotent.

> So at this point it seems unknowable whether the stem cells are totipotent. Dr. Thomson said it was "not clear if they have properties identical to the inner cell mass." Dr. Gearhart, whose cells were derived from a different tissue, the embryonic germ cells, but are probably equivalent to Dr. Thomson's, said he thought they would probably prove totipotent if the test were permissible. (Wade 1998)

In their *Science* article, Thomson and co-workers referred to the cell lines as pluripotent, while making the point that the totipotency test in other mammals – germ line transmission – could not be carried out with the human cell lines. They did however highlight that the human cells could form trophoblast as well as cells from all three germ layers (Thomson et al. 1998).[2]

In another article in the *New York Times*, various stakeholders in the debate on the status of human embryos were sounded out. A geneticist from Princeton University, Lee Silver, expressed his annoyance at what he essentially took to be semantic – as opposed to scientific – distinctions.

> Nonsense [...] If what matters, as the Government lawyers wrote, is "the capacity to develop into a human being, then human embryonic stem cells are the moral equivalent of embryos. Metaphysically, it's all the same", said Silver, who complained that while he was in favour of embryonic stem cell research, he was offended by what he referred to as 'the winking and nodding of scientists who do not want to admit the potential of the cells to become babies'. (Kolata 1999)

Dr. Silver's technical position on the prospective developmental capacity of stem cells was shared by Canadian scientists prominent in the stem cell field, who six years earlier had published a collaborative paper that described how they had managed to produce completely embryonic stem cell-derived mice from a mouse embryonic stem cell line (Nagy and Rossant 1993). 'I don't think there's

---

2   For hESC to be considered pluripotent it must, at a minimum, develop into cells from all the three germ layers. The formation of trophoblast is considered to be a marker for totipotency. Understood in a UK context and set in the present day, the distinction between totipotency and pluripotency is not particularly politically charged. Over the course of our research project, the notion that human ES cells show signs of 'totipotency' has been repeatedly confirmed in informal conversations with scientists. One embryologist summarised the position as follows: 'What we don't know can't hurt us and it rather depends what you go looking for, doesn't it?' (Fieldnotes 2007a). Sager (2006) also reports how Thomson himself at a Keystone lecture in 2003, when asked about the developmental capacity of human ES cells, summarily declared that hESCs had always been able to produce trophoblast.

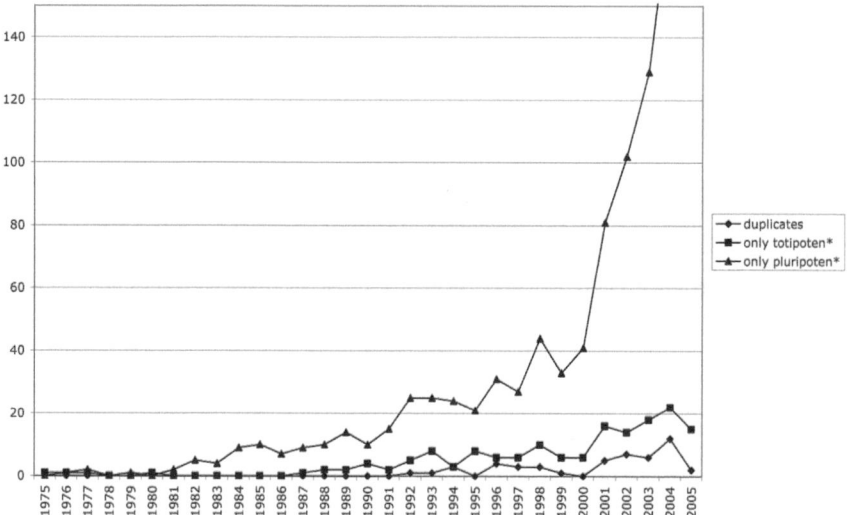

Figure 2.1   Distribution of 'pluripoten*' and 'totipoten*' in ES cell articles 1975-2005

a theoretical or practical impossibility of creating a completely stem-cell derived human being, if one wanted to do that', said one of the lead authors.

Despite such objections, the term pluripotency took hold and became an integral part of the hESC assemblage (Sager 2009). In his detailed analysis of American stem cell discourse between 1998-2001, Morten Sager describes the change in terminology that took place soon after the announcement of successful derivation of human stem cells (Sager 2006). The hESCs underwent a terminological calibration that disentangled them from embryos and placed them in a separate, and politically more palatable, category. Previous to 1998, the terms totipotent and pluripotent had been used interchangeably and often synonymously in the scientific literature. After the 1998 congressional hearings the preference for the term 'pluripotency' increased exponentially.

Pluripotency as a defining and demarcating trait of the hESC bio-object was thus originally articulated in a specific regulatory and political context. This specific articulation has subsequently presented the hESC community with a conundrum when the cells increasingly moved into a different regulatory zone. Before turning the attention to this shift however, the paper will briefly illustrate how pluripotency née 1998 protected the hESC from a potential competitor – the adult stem cell.

Higher potency – demarcating the embryonic

Pluripotency has also worked as a supporting pillar for the argument that embryonic and somatic stem cell sciences need to run concurrently and that one cannot replace the other. Such an argument has been important to make for scientists working with embryonic stem cells, to counter suggestions that there are other more suitable sources from which one can procure stem cells.

While scientists are often at pains to point out that there is no rivalry between the fields as such, nevertheless the existence and expansion of adult stem cell research remains in a competitive relationship to the embryonic field. Critics of embryonic stem cell research frequently cite adult stem cells both as a promissory way forward and an ethical rescue plan; according to this position adult stem cells have the potential to provide cures for degenerative diseases without having to destroy an embryo in the process. In this context, pluripotency is no longer used to underline the limitations of hESCs but instead their superior developmental capacity to adult stem cells, the latter being only multipotent. Multipotent stem cells' potential are restricted to particular lineages; so for example can a hematopoetic stem cell differentiate into any type of blood cell but could not become a neurone. This model of adult stem cells as restricted to specific lineages was however challenged when some scientists started to report anomalous findings suggesting that adult cells under some conditions might be able to break through a tissue-specific barrier and, as it were, re-differentiate into a different cell type. Such claims, if validated, would imperil the existing if somewhat unstable equilibrium built upon pluripotency as the main demarcation between embryonic and adult stem cells, and set the fields in open competition.

Plasticity

The first adult stem cells to be isolated, characterised and used clinically were hematopoetic stem cells (HSC) and most that is known about adult stem cells come from HSC studies (Raff 2003; Kraft 2009; Brown et al. 2006). Around the turn of the century, a number of papers made claims regarding so-called adult stem cell plasticity, and most of the plasticity studies used HSCs or cells derived from bone marrow (BM). Ferrari et al. (1998) received a lot of attention when they reported in *Science* that mouse bone marrow (BM) cells could give rise to skeletal muscle cells when transplanted into a damaged mouse muscle. This was followed by a series of papers reporting that transplanted BM cells could produce many cell types beyond their normal lineage if put into a new environment (Petersen et al. 1999; Theise et al. 2000; Lin et al. 2000; Orlic et al. 2001; Brazelton et al. 2000; Mezey et al. 2000). Furthermore, HSCs were reported to produce epithelial cells of the liver, gut, lung, and skin (Krause et al. 2001). When neural stem cells (NSC) from mouse brain were claimed to form blood cells in a paper provocatively named "Turning brain into blood" (Bjornson et al. 1999) it was received with a degree of bafflement, as were reports that the NSCs could also produce skeletal muscle cells (Galli et al. 2000) and that when the cells were injected into an early-stage blastocyst it would give rise to cells from all germ layers (Clarke et al. 2000).

The most startling aspect of some of the plasticity reports were the suggestion that cells originating from one germ layer could shift sideways not only to 'neighbouring' cell types but transform into a cell type of a different germ layer. This apparent questioning of some of the central tenets of cell development and differentiation was therefore a double-edged sword, threatening not only the developmental superiority of embryonic stem cells, but founding principles of the HSC field itself as well. The plasticity claims were met with a variety of reactions among scientists working with adult stem cells ranging from outright rejection, over cautious scepticism, to predictions that plasticity would reposition fields such as HSC research at the forefront of regenerative medicine. Regardless of the mixed responses, the potential promise of adult stem cell plasticity did result in new research funding being funnelled into the HSC field in particular and the inclusion of 'plasticity sections' in leading cell biology journals were testimony to this new prominence.

Indicative of how controversial the plasticity claims were, was the publication of negative results in high-profile journals such as *Nature* (Morshead et al. 2002) and *Science* (Castro et al. 2002), and the formulation of unusually rigorous (some said rigid) standards in the design, execution and interpretation of plasticity experiments that critics demanded must be adhered to for results to be considered valid (D. Anderson et al. 2001). In the period between 2005 and 2007, 9 out of the 20 most-cited papers on the topic of stem cells dealt explicitly with plasticity in adult stem cells, and a further three were frequently cited in that debate (Essential Science Indicators 2007). According to a feature writer for *Science*, the plasticity claims instilled the same passions in the field of stem cell research as had claims of cold fusion in physics: 'In hallways during conference coffee breaks, in strongly worded journal commentaries, and in behind-the-scenes conversations, the stem cell community is picking apart, and sometimes battling over, the evidence: Can cells from one type of tissue be induced to look and act like cells from a different tissue?' (Holden and Vogel 2002).

The intense debate that followed the plasticity and transdifferentiation reports illustrates how necessary pluripotency was not only as a defining, but as an *exclusive* trait in hESCs. Transdifferentiating adult cells with the ability to transcend germ layer boundaries could make claims to the same potential as their embryonic colleagues but without being weighed down by the ethical baggage, thus seriously undermining the careful discursive and material arrangement of the hESC bio-object – the pluripotent human embryonic stem cell.

Standard markers – characterising stem cells

Why then, one may ask (and scientists increasingly do), does it matter whether a stem cell line can make a 100 or 200 different cell types? One of the main discussion points has been how to, and in later years whether one really ought to, establish 'true pluripotency' in the cells. Pluripotency is, as mentioned previously, a state that is difficult to maintain. In fact, it is difficult to maintain the cells undifferentiated, let alone to know for sure that they still hold within them the capacity to become every type of human cell. The only certain way of testing a future projection is to usher in that future, and the gold standard for pluripotency is therefore to grow a teratoma – an encapsulated tumour consisting of tissue from all three germ layers. In simpler but less palatable terms, to grow a tissue ball of hair, teeth, nerves, skin and so on and then use histology to ascertain that 'representatives' for all the different cell types are present. This is done by injecting undifferentiated cells into a mouse, and the test is thus both costly and assumes a particular infrastructure that is not available to all laboratories. It requires an animal license together with all the materials and procedures that need to be in place to get such approval, as well as in-house histology expertise or alternatively established links with experts; a university laboratory might for example turn to the local immunology department. Teratomas are therefore in practice rarely used to test for pluripotency, apart from when a new line is announced (but not always) or in cases where the authenticity of a line has been questioned.

The International Stem Cell Initiative (ISCI) has, alongside its sister banking project (ISCBI), attempted to arrive at some consensual agreement of what a workable or 'reasonable' definition of pluripotency might be. The gold standard test – to grow a teratoma – demands considerable skills and resources both in terms of cell lines and labs. In the context of 'doable banking' therefore, pluripotency must remain a conjecture and be expressed through a series of genetic and cell surface markers rather than be ushered into a differentiated future in which a line's 'true' pluripotency has taken a material form (Webster and Eriksson 2008). When asked in an invitee-only meeting whether growing teratomas ought to be a criterion for banking a stem cell line, participants balked. Apart from being costly and time-consuming for all, such a criterion was thought to exclude countries with developing economies, which would in turn undermine the legitimacy and usefulness of ISCI and indeed the entire International Stem Cell Forum (Fieldnotes 2007c). Informally scientists also expressed concern that their own lines might not 'make it' and that a quest for the grail of true pluripotency might not be in anybody's interest, as full potential is of theoretical rather than practical value. 'It all depends on what one wants to do with the cells' was a generic statement from scientists when a conversation turned to pluripotency (Fieldnotes 2007b). Another way of putting this would be that while 'true pluripotency' is relevant to the public hESC articulation, the materiality of the hESC as found in the laboratory combined with the infrastructure of stem cell research both internationally and at a local level, calls for a different configuration of pluripotency.

Markers and materiality

The gold standard teratoma test is thus comparatively rare, and instead different laboratories favour different markers in their routine checks, although some markers have over time become nearly ubiquitous in their spread and now form an obligatory passage point for anyone wishing to publish results from their hESC work (Latour 1987). Often depending on their original academic background, stem cell scientists have different preferences for different markers and such preferences are in turn linked to technological skills and familiarity with specific methods of analysis. In their routine work most scientists and technicians working with hESCs will use just a few markers to ascertain the cells' continued undifferentiated state, together with daily 'eyeballing'.[3] The International Stem Cell Initiative, and other organisations making concerted efforts to standardise the field, are however attempting to establish a standard panel of markers to characterise hESCs.

In the first of the ISCI projects, laboratories were invited to run a number of assays on their stem cell lines and report the results back. The overall idea was to establish a range of possible 'stem cell signatures' and ultimately to establish what all lines have in common, what differences between them that are mere artifacts of local lab practice, and what differences that might be inherent in the original material. Despite the experiments being designed with the utmost simplicity in mind, the work turned out to be very demanding of laboratories' time and resources and the analysis of the results proved very complex.

Some research groups struggled with growing the quantities of cells needed to run the tests and complained that not only had the study been designed with a large laboratory in mind, but the quantities required for the tests presupposed particular methods of growing the cells. Around the time for the first ISCI meeting there was a general suspicion that chromosomal changes in lines who had been grown over a long period of time could be connected with the way in which they were being passaged, that is moved from one vessel to be 're-potted' in the next. Some groups therefore preferred to carefully scrape the cells by hand rather than to use an enzyme to detach them, as the latter treatment was thought to be harsher. Others were concerned that enzymatic passaging of the cells would affect the analysis, as such passaging tended to generate colonies with a higher proportion of differentiated cells that would therefore skew the results.

This complicating factor was brought up with some regularity during the first invitee-only ISCI meeting, namely that the characterisation work built on a form of circular reasoning. In order to establish markers for pluripotent hESCs, research groups were asked to examine the lines they assumed to be pluripotent hESCs and report their various characteristics. These characteristics were then compiled and compared and formed the basis for a standard hESC signature. Remarks regarding circularity made by some participants were met with a certain amount of frustration, in the words of one influential NIH represesentative 'a stem cell is a stem cell is a stem cell' (Fieldnotes 2005). Another and perhaps more interesting argument was that such a 'bottom-up' approach was the only appropriate one to develop 'robust' standards.

Another frequently expressed concern during the same meeting was the risk that results from different cell lines were confounded by the laboratory they came from. As different laboratories favour different culture systems, and the cells adapt to their environment, the differences recorded could just as well be a

---

3　Undifferentiated hESCs have a particular look, and training one's eye to see how they are faring and whether they are at risk of differentiating is an important skill.

'lab difference' as a 'cell difference'. Furthermore, the differences – even when consistently expressed and apparent in more than one laboratory – might be the result of which cells were selected for passaging. In everyday hESC practice, technicians will determine which cells that look 'healthy and happy' from any one culture dish and re-plate them. The concern was, then, that the cells that did well under a laboratory regime were the ones that had adapted a bit too well. In the words of one scientist, 'perhaps we ought to take the ones who look a bit ropey, perhaps they're the real stem cells' (Scientist T 2005). This latter worry also tied to the circularity argument – what if, said some scientists, the standard signature was based on cells that were doing well for a bad reason?

In the early stages of the Initiative one of the main talking points was thus the inability to distinguish 'noise' from 'real results' and, importantly, the risk of creating a skewed standard signature on the basis of such data. Such a signature would also risk being self-affirming due to the cells' tendency to adapt to any given culture system. A standard characterisation of a pluripotent 'hESC proper' detailing particular mandatory markers, combined with a standardisation of how to grow the cells, would thus have material and as well as ontological consequences.

The ISCI project collated, compared, sifted through and formalised local know-how in order to form the basis for a collective or universal statement regarding the character of hESCs. This was seen as an important step politically as well as scientifically; in the words of the NIH representative 'it is crucial to have a public domain view of what a stem cell line is and what the markers are'. This tension between a will to maintain a provisionary and critical attitude to both data and its interpretation on the one hand, and a need to mark stability and thus viability towards an 'outside' was a recurring feature of the discussions in the first ISCI meeting.

Pragmatic pluripotency – less pluri but more potent

When the field of human embryonic stem cell research moves from proof of principle towards pragmatics – that is from the ubiquitous discursive 'bench' towards clinical application – 'true' pluripotency begins to be reconsidered. As the prospect of clinical application has become increasingly discussed, pluripotency in hESCs has gone from being an almost axiomatic trait whose underlying regulatory mechanisms need to be meticulously explored and harnessed, to become a concept that is problematised both practically and semantically in scientific meetings and workshops. For purposes of clinical applications, pluripotent stem cells are less attractive: unlimited potential can also mean limited certainty. This regulatory logic has resulted in a gradual change of emphasis in the field, from possibility to function. Here, pluripotency is radically reconfigured and slotted into a 'niche model' that stresses the practical use of lines instead of their inherent potential.

The same hESC configuration that was previously a valuable discursive resource, now threatens to pose a practical problem. Unless this boundless potentiality can be harnessed, the truly pluripotent cell will be of little or no value.

This would certainly be the case in regard to the field of regenerative medicine; in fact, endless possible differentiation paths could equally and easily be construed as endless potential risk. Thus, cell lines that have previously been construed as 'stubborn' in their preference for a particular developmental pathway, would in a clinical setting be much preferable to cell lines displaying 'true pluripotency'. Scientists working in the field have increasingly begun to call for a sharpening of terminology; references to pluripotency could and should in some instances usefully be replaced by references to the cells' undifferentiated state, thereby not making claims to everything they could become but only to their current state of developmental suspension. While perhaps seemingly a mere technicality, this represents a major shift in how hESCs are talked about and understood. A hESC with, as it were, a limited pluripotency would have been an oxymoron around the millennia. As the hESC bio-object has travelled from principle towards practice, the attitude towards its pluripotency has taken on a novel and pragmatic character, in the words of one scientist: 'Ideally, in the clinic, you'd want unipotent stem cells' (Fieldnotes 2007a).

The regulatory logic that governs a clinical context requires reliable repetition, and such reliability can in some instances compensate for epistemic uncertainty in regard to the underlying mechanism (Hogle 2009). In the case of hESCs, much work remains before science can make claims to have charted and understood the many and varying differentiating pathways. A reconfiguration by which the focus shifts from conjecture to outcome would considerably narrow the scope for the pluripotent hESC, or ultimately render superfluous pluripotency as a central feature of the hESC bio-object.

Configurations of a bio-object

Bio-objects are embodiments of knowledge-in-the-making that challenge existing taxonomies. The concept captures the idea of a material-epistemic vessel that navigates and transcends different boundaries and contexts, with ontic consequences – for example laboratory, banking, clinical or political arenas. The hESC is by definition in ontological suspension or transition; its object-hood is, as it were, even more loosely tied to its presumed or perceived ontology than is perhaps usually the case. It is considered to have the potential to become new life both generally (metaphysically) and specifically (to become a neurone or a cardiomyocyte or some other cell). The form of its being hinges on a web of actors, political considerations (both internal and external politics), practices, material and meaning.

The pluripotency of hESCs represents the limits that are drawn around the 'becoming' – making the cells a) ethically/politically more savoury and b) a more promising, or as promising, an investment as adult stem cells. However, as the context changes this definition risks excluding players that the Anglo-European stem cell community need to keep on board; if certain tests or markers for pluripotency are made mandatory in banking then some laboratories or even countries might not be part of the international banking project. To complicate matters further, the fact that the cells adapt to their immediate physical surroundings make them liable to respond materially to a standardisation protocol that is in part discursively driven. This highlights the tension between the objects of science and what is (not) known about them, what Rheinberger (1997) calls their preliminarity. Such a preliminary bio-object rhymes beautifully with the ontological shaping of the not-yet-known stem cells through epistemic practices, such as characterising and creating a stem cell 'signature'.

References

Anderson, D., Gage, F. and Weissman, I. 2001. Can stem cells cross lineage boundaries? *Nature Medicine*, 7, 393.

Bjornson, C. et al. 1999. Turning brain into blood: A hematopoietic fate adopted by adult neural stem cells in vivo. *Science*, 283, 534.

Brazelton, T. et al. 2000. From marrow to brain: Expression of neuronal phenotypes in adult mice. *Science*, 290, 1775.

Brown, N., Kraft, A. and Martin, P. 2006. The promissory pasts of blood stem cells. *BioSocieties*, 1(3), 329-48.

Castro, R. et al. 2002. Failure of bone marrow cells to transdifferentiate into neural cells in vivo. *Science*, 297, 1299.

Clarke, D. et al. 2000. Generalized potential of adult neural stem cells. *Science*, 288, 1660.

Dickey-Wicker 1996. Dickey-Wicker Amendment. I. No. 104-99, paragraph 128, 110 Statute 34. 1-26-1996.

Eriksson, L. and Webster, A. 2008. Standardizing the unknown: Practicable pluripotency as doable futures. *Science as Culture*, 17(1), 57.

Essential Science Indicators, http://esi-topics.com/stemcells2007/papers/map.html [accessed 23 May 2008].

Ferrari, G. et al. 1998. Muscle regeneration by bone marrow-derived myogenic progenitors. *Science*, 279, 1528.

Fieldnotes, 2005. 1st International Human ES Cell Workshop, Bar Harbor.

Fieldnotes, 2007a. 2nd UK Stem Cell Meeting, Imperial College London.

Fieldnotes, 2007b. 3rd International Human ES Cell Workshop, Bar Harbor.

Fieldnotes, 2007c. ISCF Cell Banking Meeting, Bar Harbor.

Galli, R. et al. 2000. Skeletal myogenic potential of human and mouse neural stem cells. *Nature Neuroscience*, 3, 986.

Hogle, L.F. 2009. Pragmatic Objectivity and the standardization of engineered tissues. *Social Studies of Science*, 39(5), 717-42.

Holden, C. and Vogel, G. 2002. Plasticity: Time for a Reappraisal? *Science*, 296(5576), 2126-9.

Kolata, G. 1999. When a Cell Does an Embryo's Work, a Debate Is Born. *The New York Times*. Available at: http://www.nytimes.com/1999/02/09/science/when-a-cell-does-an-embryo-s-work-a-debate-is-born.html?ref=gina_kolata [accessed 21 June 2010].

Kraft, A. 2009. Atomic Medicine: The Cold War Origins of Biological Research. *History Today*, 59(11). Available at: http://www.historytoday.com/alison-kraft/atomic-medicine-cold-war-origins-biological-research [accessed 27 January, 2011].

Krause, D. et al. 2001. Multi-organ, multi-lineage engraftment by a single bone marrow-derived stem cell. *Cell*, 105, 369.

Latour, B. 1987. *Science in Action: How to Follow Scientists and Engineers through Society*. Cambridge, MA: Harvard University Press.

Lin, Y. et al. 2000. Origins of circulating endothelial cells and endothelial outgrowth from blood. *Journal of Clinical Investigation*, 105, 71.

Mezey, E. et al. 2000. Turning blood into brain: Cells bearing neuronal antigens generated in vivo from bone marrow. *Science*, 290, 1779.

Mol, A. 2003. *The Body Multiple: Ontology in Medical Practice*. Durham, NC: Duke University Press.

Morshead, C. et al. 2002. Hematopoietic competence is a rare property of neural stem cells that may depend on genetic and epigenetic alterations. *Nature Medicine*, 8, 268.

Nagy, A. and Rossant, J. 1993. Production of completely ES cell-derived fetuses, in *Gene Targeting: A Practical Approach*, edited by I.A. Joyner. Oxford: IRL Press at Oxford University Press, 147-79.

NIH, 2006. Stem Cell Basics. Stem Cell Information [World Wide Website]. Bethesda, MD: National Institutes of Health, *U.S. Department of Health and Human Services*. Available at: http://stemcells.nih.gov.ezproxy.ub.gu.se/info/basics/defaultpage [accessed 31 January 2011].

NIH, 2011. The National Institutes of Health resource for stem cell research. *Stem Cell* Information [World Wide Website]. Bethesda, MD: National Institutes of Health, US *Department of Health and Human Services*. Available at: http://stemcells.nih.gov.ezproxy.ub.gu.se/info/media/defaultpage [accessed 14 December 2006].

Orlic, D. et al. 2001. Bone marrow cells regenerate infarcted myocardium. *Nature*, 410, 701.

Petersen, B. et al. 1999. Bone marrow as a potential source of hepatic oval cells. *Science*, 284, 1168.

Rader, K.A. 2004. *Making Mice: Standardizing Animals for American Biomedical Research, 1900-1955*. Princeton, NJ: Princeton University Press.

Raff, M. 2003. Adult Stem Cell Plasticity: Fact or Artifact? *Annual Review of Cell and Developmental Biology*, 19(1), 1-22.

Rheinberger, H. 1997. *Toward a History of Epistemic Things: Synthesizing Proteins in the Test Tube*. Palo Alto, CA: Stanford University Press.

Sager, M. 2006. *Pluripotent Circulations: Putting Actor-Network Theory to Work on Stem Cells in the USA, Prior to 2001*, Göteborg: Acta Universitatis Gothoburgensis.

Sager, M. 2009. En bildad blick på bioteknik. I *På spaning efter teknisk bildning*. Stockholm: Liber, ss. 86-108.

Scientist T, 2005. Interview. Statement of Harold Varmus, M.D., Director, National Institutes of Health, before the Senate Appropriations Subcommittee on Labor, Health and Human Services, Education and Related Agencies, 26 January 1999.

Stem Cell Basics, in *Stem Cell Information* [World Wide Web site]. Bethesda, MD: National Institutes of Health, U.S. Department of Health and Human Services, 2009 [Friday, 28 January 2011]. Available at: http://stemcells.nih.gov.ezproxy.ub.gu.se/info/basics/defaultpage.

The National Institutes of Health resource for stem cell research, in *Stem Cell Information* [World Wide Web site] (Bethesda, MD: National Institutes of Health, U.S. Department of Health and Human Services). Available at: http://stemcells.nih.gov.ezproxy.ub.gu.se/info/media/defaultpage [accessed 14 December 2006].

Theise, N. et al. 2000. Liver from bone marrow in humans. *Hepatology*, 32, 11.

Thomson, J.A. et al. 1998. Embryonic Stem Cell Lines Derived from Human Blastocysts. *Science*, 282(5391), 1145-7.

Wade, N. 1998. Primordial Cells Fuel Debate On Ethics. *The New York Times*. Available at: http://www.nytimes.com/1998/11/10/science/primordial-cells-fuel-debate-on-ethics.html?ref=nicholas_wade&pagewanted=2 [accessed 21 June 2010].

Webster, A. and Eriksson, L. 2008. Governance-by-standards in the field of stem cells: Managing uncertainty in the world of 'basic innovation'. *New Genetics and Society*, 27(2), 99.

Chapter 3

Water – An Exploration of the Boundaries of Bio-objects

Ragna Zeiss

Water accounts for about 70 % of a cell's weight, and most intracellular reactions occur in an aqueous environment. Life on Earth is thought to have begun in the ocean, and the conditions in that primeval environment put a permanent stamp on the chemistry of living things. Life therefore hinges on the properties of water. (Alberts and Bray 2004: 48)[1]

Water has numerous relations to life. It is widely seen as the source of life or, at least, as a substance essential for the origin of life. It is also commonly regarded as a prerequisite for life. Searches for life, whether on earth or other planets, concentrate on finding water on the assumption that there is no life without water. Further, water is considered to make up a large percentage of cells, organs, blood and the brain: 'Water is necessary to maintain the volumes of body compartments, for excretion of waste products and as a medium in which biochemical reactions occur' (Ahmed et al. 2007: 208). Bodies are thus viewed as largely constituted by water. Nevertheless, water is generally not perceived as a socio-technical phenomenon where new mixtures of life occur or to which 'life' is attributed – the sketch given of a bio-object in the introduction of this book.

This chapter asks how it is possible that something so essential for life and of such unquestioned fundamental importance for life is not considered subject to a process of bio-objectification. Ingredients of possible explanations discussed below are the invisibility and mundane status of water, the strong link between water and naturalness, and the paradigm of essentiality of water. It is argued that water is objectified – made into an object – as a life giving property rather than bio-objectified. Bio-objectification processes incorporate two key features absent in the case of water: they embody living objects and debates and contestations about life and living objects. By taking the case of water, this chapter starts to explore the boundaries of bio-objects and their bio-objectification.

---

1   Copyright © 2004 *From Essential Cell Biology* by Bruce Alberts et al. Reproduced by permission of Garland Science/Taylor & Francis Books, Inc.

Silenced water

Due to its importance and omnipresence (or absence in some places of the world), it seems very clear what water is. We all recognise it and know what it tastes and feels like. This may easily lead to thinking that water has always been perceived in the same way and would explain that very little discussion has taken place about water. It would be easy to assume that, in the words of Phillip Ball (2002a: x): 'water has throughout this time been just what it is today: H_2O, a remarkable chemical compound with a private history'. The philosopher Ivan Illich, on the other hand, refuses 'to assume that all waters may be reduced to H_2O'. (Illich 1986: 4) The problem with reducing water or 'stuff' as he calls it, to H_2O is that the stuff is then a-historical. Instead Illich insists on exploring the 'historicity of matter'. Such a focus on the historicity of matter illustrates that the physical and material properties of water and classifications of good water quality are neither universal nor self-evident.

First, ideas about what water is have changed over time. In ancient Greece water was considered as one of the essential elements of which the world was built up. Thales of Miletus (c.620-c.555 B.C.) believed that the physical world was constituted of only one fundamental substance: water. (Ball 2002b) However, others argued that the primary substances were air or fire. It was Empedocles (c.490-c.430 B.C.) who postulated the four elements, earth, air, fire and water that were to become so important in Western natural philosophy. Like Parmenides, he held that one kind of matter cannot become another kind of matter. The four elements were considered eternal and fundamental kinds of matter. They persisted in science at least until the end of the 17th century. Some, such as the Canadian writer Northrop Frye, argue that the elements still play an important role in modern culture: 'The four elements are not a conception of much use to modern chemistry-that is, they are not the elements of nature. But ... earth, air, water and fire are still the four elements of imaginative experience, and always will be' (Ball 2002b: 13). In the 18th century various experiments by different people led to the view that water is composed of two elements – substances that cannot be broken down into simpler substances – hydrogen and oxygen. These featured amongst the 63 elements in the Periodic Table Dmitri Mendeleev first published in 1869 and which became the basis for chemistry education. Water had become H_2O.

Second, also the quality of water has been subject to analysis. What counts as 'in' and 'out' of place (in the water) differs across cultures and time periods. (Douglas 1966, Mullin 1996) Purity and impurity of water were often classified in terms of the source of the water – some waters were seen as purer and healthier than others – and the use of water (drinking, cooking, washing, with drinking water being the highest in this hierarchy). Appearance, smell and taste of water were the guiding principles (Goubert 1989: 37). An example from early medieval Italy:

> For the senior Pliny no water was better than that drawn up from a well; but
> Columella observed that spring water was best, well water second, and cistern

water least acceptable. According to Macrobius the problems were greatest for those who drank melted snow, for this type of water had lost its health giving vapor and contained a predominance of solid, earthy elements.[2]

In mid ninth century Lupus, abbot of Ferrières, said that he only drank water in emergencies when no wine or other alcoholic beverages were available. This water was then clear, 'drawn from a well or sparkling stream, not murky water from a cistern'.[3] The importance of clear water and its relation to health is, however, not universal. A study by Jan-Olof Drangert (1993: 108) about Tanzania contends that:

> The Swahili word for clean water, *maji safi*, alludes to its physical appearance i.e. clean as opposed to *maji machafu* (dirty water). Water is not considered *safi* if it contains small creatures or smells bad, but it does not necessarily have to be clear. Clear water may be rated as less tasty, while milky water may be rated as more filling.

(Healthy) water has thus not had a stable meaning across time and cultures. Despite this, no records show that the meaning of (good) water has been subject to heated or continuous debate. It takes an anthropologist to observe or historian to compare the rare and very few descriptions of water and to conclude that these differ. Two possible explanations come to mind.

First, water has for a long time, and largely still is, a local matter. Local or historically different ideas about water quality may not be challenged if not much interaction between these ideas exists. Increasing travel, commodification of and trade in water could make these differences more noticeable, although, as shown below, ideas about clean water have currently become more standardised.

Second, very little has been written about water at all. The historian Squatriti studied different uses of water in early medieval Italy and was surprised at the very few sources he found on this subject: 'The scarcity of such evidence even over the course of the six hundred years this study addresses is most striking' (Squatriti 1998: 8). The most probable explanation for lack of sources reflecting on water, water use and attitudes towards water is that water may have been too taken for granted, too much embedded in and inseparable from daily life and therefore too invisible to be subject to explicit reflection. Squatriti indeed argues that 'the reason for the relative invisibility of these natural sources of water for domestic use is probably their banality' (Squatriti 1998: 11).

Water can be said to be largely silenced throughout history and across cultures. An investigation of, for example, whether and how water has become

---

2   Pliny, ed. G. Serbat. 1972. *Naturales Quaestiones*, 31.23: 42. Paris. Columella, ed. H. Boyd Ash and E. Foster. 1941. *Res Rustica*, 1.5.1-2: 58. London. Macrobius *Saturnalia*, 7.12.25-6: 452-3. In: Squatriti 1998: 37.

3   *Epistolae*, ed. E. Dümmler. 1902. *MGHEpistulae* V.1, 109: 94. Berlin. In: Squatriti, 1998: 36.

discursively constructed through rhetorical strategies (see the chapter by Holmberg in this book that illustrates this in respect to transgenic mice) is barely possible. Not because the discourses have been lost and archives are not accessible, but because explicit discourses around water have probably hardly existed at all. Relations between water and life may not have been contested, at least not to such an extent that they upset social, cultural and political boundaries or systems of classification around life. Water has been, in silence, presumed to be a medium for life. A careful, tentative conclusion may be drawn that water, in the historical periods touched on above, was not constructed as a bio-object – it was neither regarded as a socio-technical phenomenon nor did it have a 'life' as such.

Surfacing of water

Very rarely, (healthy) water has been subject to some discussion. One of these uncommon occasions where water quality surfaced as a topic of debate was during the Industrial Revolution in the United Kingdom.[4] As a consequence, water became a government and governance matter, increasingly subject to processes of standardisation, a substance that had to be manufactured in order to be suitable for drinking and becoming part of the human body. Manufactured water could, in theory, have upset boundaries between nature and culture and so be considered as socio-technical matter. Interestingly, the production of this socio-technical matter became framed as the production of pure water. To better understand this, we need to delve more into the Industrial Revolution and the content of the discussions that took place.

In contrast with earlier periods, many historical sources describing the condition of the water can be found in this period. The physical purity of water is generally portrayed as deteriorating very quickly during the Industrial Revolution. Charles Dickens wrote:

> It was a town of red brick, or of brick that would have been red if the smoke and ashes had allowed it; but as matters stood it was a town of unnatural red and black like the painted face of a savage. It was a town of machinery and tall chimneys, out of which interminable serpents of smoke trailed themselves for ever and ever, and never got uncoiled. It had a black canal in it, and a river that ran purple with ill-smelling dye. (Dickens 1995 [1854]: 28)

---

4   Also fluoridation of public water supplies to help prevent tooth decay in children has been subject to public debate on and off since the beginning of the 20th century. As fluoride is added to the water this has led to scientific controversy which focused, similarly to drinking water treatment, on health rather than life as such. A difference is that fluoride is added to the water whereas treatment is seen as removing contaminants.

The Industrial Revolution that took place in nineteenth century Britain had a large impact on water sources, supply and demand. For the first time a large number of people lived in cities whose population grew rapidly. The water that had supplied the cities thus far became insufficient in terms of both quantity and quality. Since industrialisation the demands for water had increased. Domestic and industrial waste was deposited in rivers which would conveniently take it away from the cities (Sheail 1997). Local supplies were insufficient as well as polluted. Often river water was regarded as so polluted as to become 'utterly unsuitable' for manufacturing purposes (Hassan 1998: 20). Domestic consumers in London complained about 'finding 'leeches' in their water, as well as 'small jumping animals that looked like shrimps', 'an oily cream', a 'fetid black deposit', etc.' (Goubert 1986: 42). The Mersey, near Warrington, was described as 'black as ink at most times, and most offensive in smell'; the Wear at Durham was 'simply a gigantic cesspool ... emitting a stench vile enough to generate a pestilence'; and Bourne, where it emptied into the Wear, was 'at times ... as yellow as ochre and as thick as glue ...' (Wohl 1983: 235-6). In 1868, a committee studying river pollution 'received a letter, the writer of which considered it useful to point out that he had written it not with ink but with water from a river in Yorkshire!'[5] Yet, both the extent of the problems and their solutions were debated. There was no (immediate) recognition that something has to be done to prevent further pollution or to actively clean the rivers.

First, on the basis of the miasmatic theory, which held that bad vapours arising from heaps of waste were sources of disease and spread through the air, there was a strong belief that filth needed to be quickly removed from the cities. The miasmatic theory had a strong appeal to people and countless reformers and public servants enthusiastically advocated it. It seemed clear that on balance dumping sewage into the rivers saved far more lives than it took in the occasional outbreak of typhoid or even the rarer cases of cholera: 'Despite the work of Snow on water-borne cholera (...) the fact remained that death rates *were* declining and so polluted water did not appear to be particularly hazardous' (Wohl 1983: 239). Only very rarely did fatalities stimulate a debate on river pollution and expedite sanitary reform. As late as 1880 not everyone was convinced of the existence of germs. In 1878 a discussion was held about the circumstances in which germs could survive. Percy Frankland, a chemist-bacteriologist, was pressed to admit that 'the cholera germ was nothing but a theory; Frankland termed it an undiscovered fact' (Hamlin 1990: 244). Someone else demanded in 1886 that 'the so-called cholera germ be put on the table before him before he would acknowledge its existence'.[6]

Second, there was no agreement on the degree to which the rivers were polluted and whether or not the self-purifying power of water was sufficient for cleaning the water. Despite a survey carried out by the General Board of Health in 1848,

---

5   Pierre Boutin. 1984. Points de repère pour une histoire de l'assainissement, Bulletin d'information, CEMAGREF, 314-15. In: Goubert 1986: 59.

6   Report of the Select Committee on Rivers Pollution (River Lea), QQ 3793-97. In: Hamlin 1990: 244.

the Metropolitan Water Supply Act did not follow the recommendations issued by the General Board of Health to extend the London water supply. Three chemistry experts who were consulted about this in 1851 thought the measure unnecessary: 'The river may reasonably be supposed to possess, in its self-purifying power, the means of recovery from amount of contaminating injury equal to what is present exposed to its higher section'.[7] They were positive that the oxidation would lead to disappearance of the contamination.

Thirdly, economic interests, the distinction between public and private rights and the difficulty of proof of pollution were arguments against control of river pollution. Many did not see any immediate benefits from the building of sewer farms as promoted by the Royal Commission on the Pollution of Rivers, founded in 1864.[8] These would require money and advanced technologies and people could loose their jobs. The river was seen the only way to get rid of waste products and controlling river pollution would harm the economy (Luckin 1986). Throughout the last quarter of the century, the slowly developing legislation to control river pollution reflected sensitivity to manufacturing interests (Wohl 1983). Another problem for the Commission was that in 1864 the pollution of rivers was considered to affect only private rights and so only individuals could sue and that was expensive. Proving who caused the nuisance was difficult.

Fourthly, the need for expertise and correct analysis of water were contested. Potability had always been a matter of common sense; this made the role and authority of chemical water analysis unsure. The lack of an agreed chemical definition of safe water set chemists free to interpret analysis as they wanted to. When the Commission on the Pollution of Rivers wanted to set some chemical standards for purity and pollution levels, this was ridiculed in Parliament (Wohl 1983). Different sets of experts would usually come to opposite conclusions – the differences were often attributed to variations in the waters themselves. (Hamlin 1990). Distrust of 'expert' scientific advisers dominated the legislative mind. In parliament many people believed that one could decide by common sense how bad the pollution of a river was.

The above illustrates the type of debates surrounding water quality around the Industrial Revolution. The relation between water and health was subject to debate how bad is the pollution, is it miasma or water that carries diseases, what would be the effects of drinking sewage-polluted water – and scientists could not provide the solution. People were not sure what sort of a problem public water supply was and whose responsibility it was.

This changed in the second half of the nineteenth century when 'a society in which chemistry informed decisions in matters of health and industry was vastly superior to one in which it did not.' (Hamlin 1990: 68). Earlier some believed in a

---

7   *Report on the Chemical Quality of the Supply of Water to the Metropolis,* B.P.P. (1851), XXIII, pp. 9-10. In: Goubert 1986: 46.

8   According to Hamlin (1990) some people argue it was the political will that was lacking, whereas others suggest that the proper technology was lacking.

strong connection between drinking water and disease. Dr William Lambe stated in 1828 that water from the Thames supplied to the Londoners was 'laden with organic matter in a state of composition' which was sometimes perceptible but always influenced the body, in manners they did not know (Hamlin 1990: 73). By early 1880 most water analysts accepted that bacteria or similar organisms were the causes of water-borne disease (Hamlin 1990). Koch's research in bacteriology 'tended to replace the old filth-theory of corrupting emanations with a new germ-theory that seemed to explain the appearance of specific diseases. Instead of contact with foul airs, bodily invasion by microbes was the thing to be avoided' (Illich 1986: 74-5). Slowly, drinking water quality started to be dictated by a focus on the chemical and bacteriological composition of the water. Common sense could no longer be used to judge the quality of the water. While it was believed that the contamination with organic matter would often disappear through processes of self-purification, it could also be made to disappear through storage, filtration, precipitation, or settling and decantation (Hamlin 1990). Gradually, treatment of water became more important.

Water became, perhaps for the first time, a matter of government and governance. The modern treatment works, pipelines and laboratories, attention for public health and the local, national, continental and even global drinking water quality standards that slowly emerged as a consequence of this, have managed an increasing control over water. Water was no longer believed to be a principle – 'a set of watery characteristics to which various ethereal spirits might annex themselves' (Hamlin 1990: 77) – in relation to which it made sense to think about a variety of waters with various qualities and uses rather than waters of greater or lesser purity. Rather, water had become a hydrogen-oxygen compound and could no longer be assessed by lay people on basis of its appearance, taste and smell. Previous varieties of water and water preferences were replaced by a similar treatment standard for all waters and human beings. Currently, a very elaborate system of water companies, laboratories, treatment plants and regulatory agencies are in place in the western world to ensure that the drinking water is of sufficient quality to become part of the human body.

It could be argued that drinking water has become a materially new substance that cannot be found 'out there' in nature. The hybridity of manufactured water was generative of a large number of regulatory provisions and requirements associated with producing good quality water. Water became an object of science and a whole new set of practices and professions came into play. One could argue that health became governed already outside the body; public health was managed through the manufacturing of water. Manufactured water suggests a border crossing between the natural and the artificial as well as bodily boundaries and may shift or upset such boundaries. Why then, does the manufacturing of water not classify as a bio-objectification process?

The new practices were and are generally not seen as disrupting or invoking new conceptualisations of relations between bodies, nature and manufacturing or new mixtures of relations to life. On the basis of chemical and biological sciences, the relation between water, illness and sometimes death became more firmly established. Whereas the crossing of bodily boundaries by biological or manufactured material – DNA going from bodies into biobanks, brain chips entering the body – often gives rise to (ethical) discussions, manufactured water is seen as 'naturally flowing' inside and outside the body.

The hybridity of manufactured water has been hidden and framed instead as the production of pure water, as though manufactured water were a natural object. It is the naturalness of water that became reinforced. Water was objectified – made into a (manufactured) object – as a life giving property in relation to general population and health concerns. Not only was it seen as a medium for life – why treat the water otherwise – but also as a vector for life-threatening diseases (germs). Water was seen as 'laden with' organic matter and a 'means of transport' for harmful substances and microbes and had to be 'treated' to be restored to its natural, pure state. The manufacturing process came to be seen as the removal of pollutants and the restoration of uncontaminated water. The objectification of water became framed as a natural process: manufactured water is more natural and pure than the water 'out there'.

Molecularisation and the paradigm of essentiality

Whereas water is mostly silenced or seen as normal(ised) practice, we have seen how it surfaces in discussions once in a while. If conceptualised differently water could have been seen in terms of upsetting boundaries between nature and culture, however, these boundaries were reaffirmed instead and water was silenced again. I wish to discuss one more occasion where water could potentially be seen as upsetting or redefining boundaries – those between life and non-life – but – again – does not.

Rose (2001: 17) and others argue that 'natural life can no longer serve as the ground or norm against which a politics of life may be judged'. Rose uses the 'politics of health' of the eighteenth and nineteenth century as a setting against which he can show the politics of life itself of our own century:

> Previously it seemed that life inhered in the inescapable natural workings of the vital processes themselves. All medicine was able to hope for was to arrest the abnormality, to re-establish the natural vital norm and the normativity of the body that sustained it. (Rose 2001: 15-16)

In contrast, 'the vital politics of our own century look rather different' and are no longer delimited by the poles of illness and health (Rose 2006: 3). Humans are now able to change the very make-up of human beings due to molecularisation. During the last decade, the 'molecular optics' is claimed to have changed the vision of life itself:

Life is now imagined, investigated, explained, and intervened upon at a molecular level – in terms of the molecular structure of bodily components, the molecular processes of life functions, and the molecular properties of pharmaceutical products. (...) As the body becomes the subject of the molecular gaze, life is recast as a series of processes that can be accounted for and potentially re-engineered at the molecular level. (Novas and Rose 2000: 487)

Above it was suggested that manufactured water can perhaps also be seen as a re-engineering of human bodies. Yet, in the molecularisation discourses the potentially re-engineering of life is described as characteristic for the new era as opposed to the old where 'natural life', although part of a biopolitical age since the 18th century, was the standard. It is claimed that biological life is now understood through molecular biology and genetics rather than physics and chemistry. Attempts at understanding life, by life scientists as well as philosophers and sociologists of science, largely take place through a research focus on (those involved with) genetic material. This is based on a 'specific paradigmatic view: DNA is the "basic" principle of Life' (Salvi 2002: 28). Modern genetics has by now provided many grounds for believing that DNA is not a stable and coherent structure and that genes are much more complex than was assumed when the Human Genome Project started. Genes cannot simply be translated into "the language of life" that contains "the digital instructions" that make us what we are'. (Rose 2011: 14) Instead, many have started to speak about the era of post-genomics where the genomic body should be understood in terms of its '*displacement* within wider molecular fields' (Italics in original) (Braun 2007: 7).

Water hardly surfaces in molecular biology. It seems entirely absent in social science and philosophy scholarly work engaging with (molecularisation of) life and has a background position in the curricula, handbooks and research of life scientists. Yet, perhaps, arguments could be made for a more important role of water. Rather than solely being a medium for life and a vector for life-threatening diseases, on the molecular level, it may be possible to conclude that water constitutes life as much as proteins and DNA.[9] Although the biologist Neil Campbell does not explicitly refer to water, when discussing atoms, molecules and chemical bonds in his famous textbook 'Biology', he states:

The phenomenon we call life is the cumulative product of interactions among the many kinds of chemical substances that make up the cell of an organism (...) Somewhere in the transition from molecules to cells, we would cross the blurry boundary between nonlife and life. Life emerges from the integrated organization of the whole organism. (Campbell 1990: 20)

---

9   At the same time, the Jewish perspective on stem cell research argues that 'during the first forty days of gestation, their status [genetic materials, embryos] is, according to the Babylonian Talmud, 'as if they were simply water' (Dorff 2001: 91; see also Prainsack, 2006; Zoloth, 2001). The large relative quantity of water in embryos makes them less/not alive according to this perspective.

On this level, the boundary between life and non-life blurs: life results from or may be manufactured with help of molecules which in themselves are not characterised as living. As the body becomes the subject of the molecular gaze, it can be argued that not only genetics, but also water deserves attention as part of (constituting) life. In fact, the connection between water and life, not so established in earlier periods as the chapter has shown, may be reconfigured at this molecular level.

For its important function of shaping the form and function of proteins, water is described as 'the medium for life' (Phillips, Kondev and Theriot 2009: 327), 'the medium of biochemical life', water serving 'as a kind of aether for biochemical action' (Phillips 2009: 328) and 'the envelope of water around its [the cell's] molecules'. (Ball 2002a: 236) This function of water is confirmed in an interview with a biochemist from Maastricht University:

> Proteins are the 'motors of the body'. DNA is the code, but needs to be translated in proteins, otherwise nothing happens. Proteins need the water to carry out their function and form the right structure. Function and structure are strongly related. We study these standard purified proteins in their function, they are soluble in water, and would not have a function without the water. (...) From an evolutionary perspective, the proteins and enzymes are made in such a way that they work in water: the functionality of water is very much intertwined with our naturally produced proteins. So, water is very important. (Interview biochemist, 11 March 2010)

Words such as 'medium' and 'envelope' seem to describe the environment for what is seen as central to life: proteins and DNA. In the words of Phillip Ball (2002a: 231) 'it is all too tempting to regard the relationship of this fluid to the biomolecules it contains ... as a carrier, a bland background on which the important business is displayed'. Although it is recognised that water is not a neutral medium – 'proteins would not have a function without the water' – it may be possible to take it a step further. Water might be 'spiced with proteins and DNA, sugars, salts, fatty acids, seething with hormones', but

> ... this won't do. Water plays an active role in the life of the cell, to the extent that we can consider water itself to be a kind of biomolecule. Without it, other biomolecules would not only be left stranded and immobile (...) – they might no longer truly be biomolecules, unravelling or seizing up and losing their biological function in the process. (Ball 2002a: 231-2)

It could be thus contended that biology and life are all about the interactions of such molecules in and with water. (Ball 2002a: 232) Many questions remain about the role and function of water. It is as yet, for example, unclear how the solvent properties of water influence the character and behaviour of the solutes as well as how these solutes modify the ability of water to act as a solvent. Amiry-Moghaddam and Ottersen (2003) state that brain function is closely linked to water homeostasis.

One of the most fundamental questions in biology, according to them, is how water is transported across cell membranes given that water transport is involved in all secretory and absorptive processes. Others argue that the fundamental question is what the structure of water in the cell is: 'This is one of the most important unresolved issues in biology. ... A good way to start an argument amongst cell biologists is to ask them what cell water is like.' (Ball 2002a: 234, 236).

Yet, despite its importance and unresolved questions, water is not highlighted as a topic to study in relation to biology and molecular life sciences. Whereas one may be able to imagine a biology handbook centred around water – water and the origin of life, water and the environment, water and plant and animal life, chemistry of water, water and the cell – my search for such a(n interesting) book remained without result. Generally, university courses and biology textbooks concentrate on cells, genes, evolution and ecology (Campbell 1990, Maastricht University Bachelor Programme Biomedical Sciences). Water is not viewed as a similar building block and is touched upon mainly in its function as a solvent. Its fluid character is resembled in the way it occasionally surfaces in biology textbooks in relation to the functioning of the other building blocks.

The biochemist mentioned earlier explained that he and his colleagues, to carry out their research, always assume that water is present. Water is seen as the necessary, although often invisible and taken for granted background and undercurrent in understanding genes, cells and organisms:

> When scientists publish models of biological molecules in journals, they usually draw their models in bright colors and place them against a plain, black background. We now know that the background in which these molecules exist – water – is just as important as they are. (Ball 2002a: 232)

Despite the potential contribution of water to the blurring of the boundaries between life and non-life, this has not raised discussion. As in the previous sections, water is clearly presumed to be essential for life – the paradigm of essentiality – but there seems no need to change the conception of water as a medium and explore it explicitly as a molecular component of life. Water remains seen as non-living material which is crucial for human bodies to exist and the medium for life rather than being life itself.

Boundaries of bio-objects

In the case of water, life and its boundaries, as discussed in other chapters of this book and by other social science scholars (see for example, Bauer and Wahlberg 2009), were never at stake as a contested category. Water has been largely silenced and taken for granted as essential for life and where it surfaces in discussions – related to pollution, fluoridation, or lack of clean water – these centre around health and illness, how pollution can and should be defined and whose responsibility it is.

Despite different conceptualisations of water over time and across cultures and even in cases where it could be argued that water crosses established boundaries, between nature and culture, inside and outside the body, and life and non-life, conceptions of water as natural and non-living material have only been re-established. Water is seen as (belonging to) normal life, similarly to Thomas Kuhn's normal science (Kuhn 1996 [1962]) whereas for example transgenic species may be seen as belonging to revolutionary life – new ways of thinking and practices are developed around what such life forms are and how they can and should be regulated.

Water is not seen as subject to a process of bio-objectification – it has neither been viewed as a living object nor has it been a central element in contestations about life forms. Yet, the concept of (bio-)objectification has helped to find alternative ways to perceive (the history of) water – as being manufactured, as constituting life. At the same time, the case of water raises questions that help to further explore the (boundaries of the) concepts of bio-object and bio-objectification.

First, the case of water shows that classifications and categories may, in some cases, be relatively fixed and stable and difficult to change or cross. New practices have been fitted into old taxonomies. Existing affordances are maintained rather than new ones opened up (see Introduction to this book). Can such examples also be found in the realm of bio-objects or do these by definition open up new affordances and have 'considerable fluidity and mobility across different socio-technical domains or arenas' (see Introduction to this book)?

Second, to what extent are the concepts of bio-object and bio-objectification related to current (public) discourses about life? Should they been seen as contemporary concepts applied to contestations of our contemporary understanding of 'life' and its boundaries only? Or alternatively, can we find examples of bio-objects in the past? Can we look back at history and apply a concept developed on the basis of situations in the 21st century? What if discourses about life were different in the past – not phrased in terms of biology?

This leads to the third question. To what kind and conceptualisation of life do the concepts of bio-object and bio-objectification refer? To what extent can it be said that bio-objects only feature in relation to the domain of the biosciences? Rose (2006: 4) argues that 'it is now at the molecular level that human life is understood' and also the introduction of this book states that we are witnessing 'that "life" as an *object* of research, intervention and innovation is increasingly represented through an idiom of science and its unquestionable regime of truth'. Yet, the case of water shows that different truth regimes co-exist and not all relations to life fall within the current truth regime of contemporary biosciences. Such other relations to life are certainly given less consideration in current social science and biomedical discourses, but do they automatically fall outside the domain of bio-objectification? And perhaps more important, are they less interesting or significant? Is it only worth speaking about life when it is contested?

The absence of contestations of life as we have seen in the case of water does not imply a negation of a relation between water and life. On the contrary, water has numerous relations to life, as a condition for life, a life giving property, an essential part of living beings. Its relations to life are celebrated, much more than relations to life of many bio-objects: 'Water, thou hast no taste, no color, no odor, canst not be defined, art relished while ever mysterious. Not necessary to life, but rather life itself, thou fillest us with a gratification that exceeds the delight of the senses' (Antoine de Saint-Exupéry 2000 [1939]: 101). Water has been central to the objectification of life as an external relation – a medium for life and a vector for life-threatening diseases, – as a central concern in managing forms of life – a governance matter, and perhaps as an internal (molecular) component of life.

Exploring the variety of relations to life is key to our understanding(s) of life. Water is especially interesting in this respect. In past and current discourses it is mostly a component of life rather than life itself as defined in Rose's terms. (Rose 2001, 2006). Yet, at the same time it demonstrates the limits of empirically attributing life to anything that lacks water at molecular level as the current epistemological paradigm of biology pre-supposes relations between water and life (on earth and elsewhere): it is life itself in the words of Saint Exupéry (2000 [1939]: 101). Already since ancient Greece, water has been considered as one of the elements constituting the earth. It can be argued that since then the relation between water and life is and has been perceived as a given, a fact and unalterable.[10] In that view, water acts as a required (epistemological) vector for successful attributions to life – central to bio-objectification processes.

Illich (1986: 25) argued that 'water remains a chaos until a creative story interprets its seeming equivocation as being the quivering ambiguity of life'. According to him (1986: 25), it is 'the deep ambiguity of the water that makes it elusive for us'. The concept of bio-object may help to creatively and alternatively interpret waters' relations to life; water may help to further investigate the (boundaries of the) concept bio-object and to address the variety of (relations to) life beyond the present time, contestations and the current truth regime of contemporary biosciences.

References

Ahmed, N., Dawson, M., Smith, C. and Wood, E. 2007. *Biology of Disease*. New York: Taylor & Francis.

Alberts, B., Bray, D., Hopkin, K., Johnson, A., Lewis, J., Raff, M., Roberts, K. and Walter, P. 2004. *Essential Cell Biology, Second Edition*. New York: Garland Science/Taylor & Francis Group.

Amiry-Moghaddam, M. and Ottersen, O.P. 2003. The molecular basis of water transport in the brain. *Nature Reviews – Neuroscience*, 4, 991-1001.

---

10 A recent – contested – study describes a bacterium from Mono Lake in California that can use arsenic instead of phosphorus as a structural building block in DNA (Matson 2010). If similar discoveries would find that water can be replaced as well, relations between water and life would need to be fully rethought.

Ball, P. 2002a. *H2O: A Biography of Water*. London: Phoenix.

Ball, P. 2002b. *The Ingredients: A Guided Tour of the Elements*. Oxford: Oxford University Press.

Bauer, S. and Wahlberg, A. 2009. *Contested Categories: Life Sciences in Society*. Farnham: Ashgate.

Braun, B. 2007. Biopolitics and the molecularization of life. *Cultural Geographies*, 14(6), 6-28.

Campbell, N.A. 1990 [1987]. *Biology*, 2nd edition. California: The Benjamin/ Cummings Publishing Company, Inc.

Dickens, C. 1995 [1854]. *Hard Times for These Times*. London: Penguin Books.

Dorff, E.N. 2001. Stem Cell Research – a Jewish Perspective, in *The Human Embryonic Stem Cell Debate – Science, Ethics, and Public Policy*, edited by S. Holland, K. Lebacqz and L. Zoloth. Cambridge, MA: MIT Press, 89-94.

Douglas, M. 1966. *Purity and Danger: An Analysis of Concepts of Pollution and Taboo*. London: Routledge and Kegan Paul.

Drangert, J.O. 1993. *Who Cares About Water? Household Water Development in Sukumaland, Tanzania*. Linköping: Linköping Studies in Arts and Science (85).

Embryonic Stem Cells, in *The Human Embryonic Stem Cell Debate – Science, Ethics, and Public Policy*, edited by S. Holland, K. Lebacqz and L. Zoloth. Cambridge, MA: MIT Press, 95-112.

Goubert, J-P. 1989. *The Conquest of Water: The Advent of Health in the Industrial Age*. Cambridge: Polity Press.

Hamlin, C. 1990. *A Science of Impurity: Water Analysis in Nineteenth Century Britain*. Bristol: Adam Hilger.

Hassan, J. 1998. *A History of Water in Modern England and Wales*. Manchester: Manchester University Press.

Illich, I. 1986. *H2O and the Waters of Forgetfulness*. London: Marion Boyars.

Kuhn, T.S. 1996 [1962]. *The Structure of Scientific Revolutions*. Chicago, IL University of Chicago Press.

Luckin, B. 1986. *Pollution and Control: A Social History of the Thames in the Nineteenth Century*. Bristol: Adam Hilger.

Matson, J. 2010. Poison Nil: Mono Lake Bacterium Exhibits Exotic Arsenic-Driven Biological Activity. *Scientific American*, 33. Available at: http://www. scientificamerican.com/article.cfm?id=arsenic-life [accessed: 21 March 2011].

Mullin, A. 1996. Purity and Pollution: Resisting the Rehabilitation of a Virtue. *Journal of the History of Ideas*, 57(3), 509-24.

Novas, C. and Rose, N. 2000. Genetic risk and the birth of the somatic individual. *Economy and Society*, 29(4), 485-513.

Phillips, R., Kondev, J. and Theriot, J. 2009. *Physical Biology of the Cell*. New York: Garland Science, Taylor & Francis Group.

Prainsack, B. 2006. 'Negotiating life': The regulation of human cloning and embryonic stem cell research in Israel. *Social Studies of Science*, 36(2), 173-205.

Rose, N. 2001. The Politics of Life Itself. *Theory, Culture & Society*, 18(6), 1-30.

Rose, N. 2006. *The Politics of Life Itself: Biomedicine, Power, and Subjectivity in the Twenty-First Century*. Princeton, NJ: Princeton University Press.

Saint-Exupéry, A. de. 2000 [1939]. *Wind, Sand, and Stars*. London: Penguin Books.

Salvi, M. 2002. *Rationalising Individuality: The Notion of Individuality in Biology, Philosophy, (Bio)ethics*. Maastricht University: PhD thesis.

Sheail, J. 1997. The sustainable management of industrial watercourses: An English historical perspective. *Environmental History*, 2, 197-215.

Squatriti, P. 1998. *Water and Society in Early Medieval Italy, AD 400-1000*. Cambridge: Cambridge University Press.

Wohl, A.S. 1983. *Endangered Lives: Public Health in Victorian Britain*. London: Methuen.

Zoloth, L. 2001. The Ethics of the Eight Day: Jewish Bioethics and Research on Human, in *The Human Embryonic Stem Cell Debate – Science, Ethics, and Public Policy*, edited by S. Holland, K. Lebacqz and L. Zoloth. Cambridge, MA: MIT Press, 89-94.

Chapter 4

Bio-objectification of Clinical Research Patients: Impacts on the Stabilization of New Medical Technologies

Conor M.W. Douglas

Patients are increasingly subject to the epistemological capture of bio-scientific processes. By this I mean that their central positioning in medical health practices is now being supplemented by increasing demands in various R&D processes. It should be noted that in many respects this is a 'voluntary capture' as patients, patient groups, and patient advocacy groups have worked hard over the past 20 years for increasing roles and responsibilities in all levels of medical R&D (Hanley et al. 2003; Brown and Zavestoski 2004; Epstein 2008; Douglas 2009). It should also be noted that the increasing demands of patients are also linked to the increasing complexity of bio-medical research in the age of genomics. Multidisciplinary research teams – such as the one described below – are now being assembled to understand the various components of disease, the interactions between genetic and environmental factors, and drug response. The novel medical science that seeks to make use of increased understanding of the role of gene-environment interactions to improve the safety and/or efficacy of new or existing drugs is known as pharmacogenetics (Webster et al. 2004). The increasing linkages between patients, medical knowledge production processes, and their associated clinical technologies reflects bio-objectification processes that are breaking-down boundaries between patients (human) and technologies (non-human), patients' status as research subject and research object, and between conventional divisions between health care and medical R&D. While these barriers are being broken-down other barriers are being built-up. Multidisciplinary R&D requires that various areas of expertise make their own disciplinary contribution to understanding the problem at hand.

As a result, participating patients are bio-objectified by these different components of the R&D team based on their epistemological and disciplinary background. First level bio-objectification occurs when the participating clinical research patient comes to be conceptualized in terms of their genetic make-up, their bio-chemistry, or the amount of money they will save the health service by staying out of hospital. This bio-objectification provides the clinical research patient a kind of fluidity that facilitates their circulation through the various components of the multidisciplinary research process. However, what this examination shows is that further – or full – bio-objectification of 'life' is not always possible, so raising problems for the stabilization of new medical technologies like pharmacogenetics (PGx). In this case we see tension between – on the one hand – the breaking down of boundaries between patient and research subject, and on the other hand the building up of boundaries between disciplines in their conception of the research participants and their biological samples. This tension results from attempts to integrate heterogeneous actors (i.e. patients) into modern medical R&D that typically requires a homogeneous and standardized clinical research patient (CRP) for the development and deployment of new medical technologies.

Once the clinical research landscape for PGx research in Britain is outlined this chapter explores those bio-objectification processes that allow for differential configurations of the CRP across the multidisciplinary research team. However, when it comes to realigning these different bio-objectifications of the CRP for the stabilization of a prospective technoscience like PGx we see levels of resistance, which suggest difficulties in fully bio-objectifying life and its inherent complexities.

The clinical research context of pharmacogenetic testing for warfarin in Britain

In 2003 the UK's Department of Health published a White Paper entitled *Our Inheritance, Our Future* that set out how the Government would go about "realising the potential of genetics in the National Health Service" (Department of Health 2003). This policy document concurred with – and acted as the foundation for – many of the leading researchers and commentators who suggested that PGx was among the most promising of the many health related benefits that would follow from the then newly 'completed' Human Genome Project (Collins and McKusick 2001; Hedgecoe and Martin 2003; Primohamed and Lewis 2004; Webster et al. 2004). The central function of this novel techno-science is to improve the safety and efficacy of new and existing drugs as well as to act as an aid in the discovery of new drugs for specific groups of people (Webster et al. 2004). Among other proposals within the White Paper was a commitment of £2.5 million "in pharmacogenetic research on *existing* medicines which patients are *commonly taking now*, or are likely to be taking soon" (2003: 64, my emphasis). One of these existing medicines being taken now by hundreds of thousands of people in the UK alone is warfarin. This is "the oral anticoagulant of choice in the UK" (Pirmohamed et al. 2004), and because it is a drug that stops the blood from clotting it is used to treat conditions such as deep venous thrombosis (i.e. blood clots in veins), pulmonary embolisms (i.e. blood clots in the lungs), as well atrial fibrillation (i.e. irregular heartbeats that can lead to clots), and other conditions like antiphospholipid antibody syndrome (i.e. blood clots in both arteries and veins that can lead to pregnancy-related complications). If left untreated these conditions can prove to be fatal by way of stroke or organ failure, and often in best-case scenarios these conditions require chronic care, supervision, and treatment.

However, because the nature of the drug is to alter the blood's ability to form clots, the recipients run the risk of internal bleeding if the wrong dose of warfarin is administered. Not only is internal bleeding seriously dangerous for the patient, but such adverse drug reactions (ADR) are also very costly for the NHS having thereby to admit these patients onto hospital wards (Pirmohamed et al. 2004; Department of Health 2003: 17). Conversely, if the wrong dose of warfarin is administered, then the patient may not be receiving the drug's full therapeutic effects, which could then lead to clotting related complications. Not only is the drug used in various routine clinical practices, but according to IMS Health – a private company providing information on pharmaceuticals – warfarin is used by "1% of the whole UK population (600,000 patients) and 6% of those over 80 years (154,000 patients) are on warfarin" (Pirmohamed et al. 2004: 1).

In light of the widespread use of warfarin to treat serious conditions, as well as the concurrent risk associated with the nature of the drug, in 2004 the Health Minister at the time – Lord Warner – announced a £842,000 project to develop a new dosing algorithm that would aid clinicians to assign patients the most suitable dose of warfarin to achieve therapeutic effects without adverse reactions. At present, it is routine practice in many Hospital Trusts in the UK to start relevant patients on what is call a 'loading dose' of warfarin. This means that before doses can be adjusted based on results from blood coagulation tests, each patient is given a high dose – often 10 milligrams – of warfarin on their first and second day of treatment, with a lowering to 5 milligrams on the third day. These first three days of treatment are followed by a return to the clinic to have a test to assess the patients' blood coagulation levels – or INR (international normalized ratio) – so that their dose can be adjusted to meet their specific needs.

This initial one-size-fits-all 'loading dose' procedure is not only potentially dangerous as patients run the risk of internal bleeding, but the concurrent levels of monitoring associated with the loading regime and the adjustments of the therapeutic to achieve acceptable coagulation levels can be seen as inefficient in terms of clinical practice, cost expenditures, and patient experience. According the 2003 White Paper "adverse drug reactions are estimated to affect around 7 per cent of patients or hospital admissions at an annual cost of about £380 million to the NHS in England alone" (Department of Health 2003: 17). Financial implications aside, it is well documented that ADRs are more dangerous in terms of mortality than other nefarious acts such as cocaine overdose and murder in Britain.[1] Because the financial burden for treating ADRs on patent-expired drugs falls on the NHS, it is in their interest to improve the safety and efficacy of this drug prescribed to

---

1   In 2005, there were 171 cocaine-related deaths in Britain (Reed, 2006) and 723 cases of homicide (Home Office, 2006). This number is dwarfed by the 1013 patients dying in Britain in 2005 from adverse drug reactions (Campbell, 2006: 15).

600,000 in England. This interest is based not only on moral grounds in terms of duty of care, but also on financial ones as well.

In light of the promise of PGx to address the inadequacies associated with the current warfarin treatment regime, the proposed new algorithm set out to take into account not only traditional factors that are known to affect the dose of warfarin that a patient should receive (i.e. age, sex, weight, diet, alcohol intake), but to incorporate an additional factor – genetic information. Ideally a PGx prescribing algorithm for warfarin would use information on all of these factors to stratify patient populations into more specific dosing categories such as: 'warfarin resistant', which would result in high than normal doses of the drug; 'warfarin sensitive, which would result is lower than normal doses of the drug; and 'normal responder', which would result in traditional dosing protocols. In attempting to develop this new prescribing algorithm a multidisciplinary team was assembled. An appreciation of the multidisciplinary nature of the PGx project is important as it will be shown how these various groups differentially configured the participating CRPs, which demonstrates the fluidity, malleability of CRPs as bio-objects.

These disciplinary components included: research nurses responsible for patient recruitment (goal for this study was N=2000) and the collection of patient data (i.e. weight, age, diet, and alcohol intake) that influences warfarin response; geneticists seeking new single nucleotide polymorphisms that affect enzymes responsible for drug metabolism as well as describing the penetrance of those genes known to have an effect on drug metabolism; pharmacologists undertaking pharmacokinetics analyses of warfarin resistance and sensitivity; biochemists monitoring patients' clotting factors in relation to their vitamin K levels (which is common and found in leafy green vegetables and fruits); different biochemists looking at clotting factor levels and at what stage the patient's blood produces clots; medical statisticians who conduct data monitoring and statistical analysis to elucidate a relationship between all of the above components, determine the relationship between the factors influencing warfarin variability, and create the sought after new algorithm; health economists to model the economic viability of the prospective algorithm and determine its cost effectiveness *vis-à-vis* the patient's quality of life and cost per serious adverse event avoided; and sociologists of science and technology to conduct social science research on the organisational barriers and facilitators to the introduction of the prospective algorithm into routine clinical practice, and to assess patient acceptance to PGx and having their dose regulated based on this kind of technology.

While strong rationale exists for the inclusion of all of these different disciplines into research for a new warfarin prescribing algorithm, what is perhaps less clear is how all of these components of the project were to fit together? Because an algorithm is essentially a decision tree in which any patient should be able to be put through by the practitioner to ascertain their safe and effective dose of warfarin, a kind of 'ideal type' or standardized patient needed to constructed on which the algorithm would be modelled.

In medical R&D these standardized patients are usually the product a large and diverse clinical research participant population whose heterogeneous biomedical information is recorded and analysed for statistical significance and generalisable outcomes. However, significant challenges exist in bringing diverse disciplines together as they conceptualize the CRP differently based on their epistemological grounding. As we shall see the circulation and trading of patient data across multidisciplinary research group is often at odds with the inherent heterogeneity within patient populations. The bio-objectification processes that *should* result in a 'standard clinical research patient' (i.e. the bio-object on which the PGx algorithm would be based) can have fundamental impacts on the stabilization of new medical technologies like PGx.

Configuring clinical research patients as bio-objects

The PGx R&D for warfarin sought to provide new guidelines that would alter clinical practice; gauge the patients' experiences of the prospective new treatment regime; derive lessons for the future of warfarin treatment; and collect information about the structure and costing of a warfarin service delivery. Each component of the multidisciplinary team of the PGx projected would require different kinds of data from the research participants. At times during their journey through the research project the CRPs would not even be considered patients, instead members of a particular discipline would treat them as 'genomes', as indication of 'vitamin K' or 'coagulation levels', or even as pounds (£) and pence. Bio-objectification can be seen as the process that allows for this abstraction and segmentation of the CRP into a set of particular configuration. The disciplinary configurations of 'the patient' are summarized below in Table 4.1, which lists the components of the study.

It is the case that all of the disciplines involved required different contributions and forms of data from the CRP who would be participating in the PGx project, but the crucial factor is that we all depended on these CRP for contributions.

By metaphorically carving up the CRP into segments of biological, chemical, genetic, and other kinds of information a reification process takes place as they become different objects of study. In the case of the PGx project, all disciplinary components were bound to the CRP in one form or another: we were all motivated to improve the safety and efficacy of warfarin for the benefit of patients on warfarin, and we all depended on contributions from the CRP to complete our respective tasks. This motivation and reliance on the patient would bind the various components of the research together for the duration of the project, but the flexibility in terms how the CRP was bio-objectified from one discipline to the next allowed each of the groups to conduct their work without conceptual infringements or confusion between the project team members. Consequently when the biochemist – for instance – would talk to the research nurses about the transportation of 'patient samples' to their laboratories across

**Table 4.1   Components of PGx warfarin project, the associated
             disciplines and concurrent bio-objectification of the clinical
             research patient (CRP)**

Component of PGx Study	Associated Discipline	Bio-Objectification of the Patient
Patient Recruitment and Patient Data	Research Nursing	CRP as reporter of lifestyle and biological information, patient as an 'n'
Genotyping/Genetic Component	Genetics	CRP as a set of genes or genome
Pharmacokinetics Analysis	Pharmacology	CRP as the biochemical site responsible for drug metabolism
Determination of Vitamin K Levels	Biochemistry	CRP as an assay in a lab
Data Monitoring and Statistical Analysis	Medical Statistics and Health Evaluation	CRP as a series of nominal data
Modelling and Cost-Effectiveness	Health Economics	CRP as costs or savings
Social Science Component	Sociology	CRP as experiential experts in warfarin treatment

the country so they could conduct their component of the research there would
be no need to enter into discussion about the people the samples belonged to
(i.e. concerns and contextualities specific to the research nurses), or what part
of the blood would be analysed (i.e. issues specific to the biochemist); rather,
work would simply get done and 'the patient samples' would be sent. Each of
the disciplinary components required flexibility of how they would bio-objectify
the CRP, and what that bio-object would ultimately contribute to the prospective
PGx algorithm. As Table 4.1 demonstrates, where the geneticist interpreted the
patient as a set of genes that might influence response to warfarin, the statistician
saw the CRP as one of the many required 'n' in a movement towards statistical
significance, and so on and so forth across the project.

While the fluidity and malleability granted through bio-objectification can
facilitate innovative research, it can also lead to knowledge production contingencies
and conflicts that can contribute to a lack of stabilization of technologies such
as PGx. For instance, CRP recruitment proved to be a major stumbling block in
the project, and that while an N=2000 was originally put forward in the proposal
the final numbers of participants were closer to N=1000. The research nurses
responsible for recruitment could have brought more CRP on warfarin into the
study, but their bio-objectification of the CRP as an 'n' were at odds with the
biochemists' configuration of the bio-object. For this group the bio-object was
not simply another participant on the way to 2000; rather they configured these
as samples, and these samples were needed from prospective CRP who had just

been recently initiated onto warfarin and also needed them to fast before blood could be drawn for the vitamin K assay. Research nurses noted that both of these factors (i.e. recent initiation on warfarin and fasting) reduced the number of patients able to be recruited as CRP, yet these factors were fundamental to the biochemists bio-objectification of the CRP. As biochemists continued to experience problems undertaking the vitamin K assay it was ultimately abandoned as a component of the project, contrary to the initial objectives that had sought to integrate this information as a vital component of the algorithm. This suggests a tension within, or resistance to, full bio-objectification of life through the course of multidisciplinary medical R&D. While the CRP needs to be differently configured across the research project, these different bio-objectifications also need to be brought together to produce the PGx prescribing algorithm.

Moreover, the differential bio-objectification of the CRP also led to conflict between the role of biochemists and the culture of statisticians. With lower recruitment numbers than anticipated, statisticians would have a harder time producing statistically relevant relationships between the various environmental, genetic, and lifestyle factors and variation in warfarin responses. The solution that the statisticians were forced to take in light of recruitment problems was to reduce CRP sample size 'required' for statistical significance. Not only did this move amount to a 'shifting of the goals posts' in terms of the original project proposal, but these problems resultant from differential bio-objectifications of CRP may have also led to conflicts with the geneticist as a smaller population meant less genetic variation from which potential genes associated with warfarin response could be found.

The prospective problem of social science integration into a prescribing algorithm can be read in a similar fashion. While the social science component of the project was concerned with creating an object of study that contained qualitative detail, depth, and breadth of life under the current – and future – warfarin treatment regimes this configuration of the CRP came into conflict with the statistician who sought representativeness of the samples of the data she sought to analyse. The qualitative nature of the social science data collected may never have been able to be integrated into the algorithm, which is essentially a decision tree that technically requires a series of well-defined successive states so that an end-state could be reached (i.e. in this case a dose of warfarin). According to Bijker and Law (1992), technologies are seen to stabilize if agreement can be reached between the relevant social actors, and 'only if the relations in which it is implicated, and of which it forms a part, are themselves stabilized' (p. 10). What is suggested here is that the bio-objectification of the CRP that allowed it to be differentially configured in contexts of care and contexts of treatment, and between all of the different components of the multidisciplinary research, was ultimately resisted by the complexity inherent to the CRP. Bio-objectification provided for the differential configurations by which the CRP could be understood by the different components of the project, but the process was not able to simultaneous provide an object that could be effectively traded and moved around the PGx research team.

Discussion and conclusion

This chapter has set out to accomplish three tasks: one was to posit that humans – in the form of CRP – can be subject to bio-objectification process through their involvement in medical R&D. The second was to display that bio-objectification processes can allow for differential configurations of the CRP across the multidisciplinary research team; and the third was to show the resistance of the CRP to move from this level of bio-objectification to one that would see the realignment of these different configurations for the stabilization of prospective techno-sciences like PGx. In describing the bio-objectifications processes that take place for the clinical research patient through the course of multidisciplinary pharmacogenetics R&D we see that the CRP was conceptualized in various ways across the project team for their respective components of the project. While the details of the various ways in which the research patient as a bio-object was differently configured across the multidisciplinary team may have been limited, what is important to note is that these differential configurations of the patient matter in accomplishing innovative, multi-disciplinary, health research goals. What this case has shown is that the malleability of the CRP that comes with the bio-objectification process can result in contingencies in the knowledge production process as the different configurations of the patient move through the various research contexts. Instead of being seen as 'blips' in the development of bio-medical technologies these problems can be seen as points of resistance to the bio-objectification process. As the differential conceptualizations of the CRP came into conflict through the course of the project the bio-objectification process began to break-down. What needed to happen for the development of a PGx algorithm is that heterogeneity of the CRP had to be closed down, and a homogenized, ideal-type, patient had to emerge. In the end the complexity of the CRP, and of the requirements of the research, meant that the components of the project could not be brought together in the form of a PGx algorithm. In the end, perhaps the abstract concept of 'life' cannot be fully reified across the bio-life sciences, or at least it cannot be done so without effects on the stabilization of new medical technologies. While other chapters in this volume explore the bio-objectification of non-human genetic resources (Tamminen), silicon cells (Vermeulen), or mice (Holmberg), when we apply this heuristic device to people – and more specifically research patients – different tensions between subject and object become central. By examining how PGx R&D differently bio-objectifies CRP, and attempts to use them as bio-objects, we are in a position to observe how epistemology configures materiality, and how materiality configures epistemologies.

References

Brown, P. and Zavestoski, S. 2004. Social movements in health: An introduction. *Sociology of Health & Illness*, 26(6), 679-94.

Campbell, A. 2006. 20 die each week after taking pills. *Metro*, 22 February, 15.

Collins, F. and McKusick, V. 2001. Implications of the human genome project for medical science. *JAMA*, 285, 540-44.

Department of Health. 2003. *Our Inheritance, Our Future: Realizing the Potential of Genetics in the NHS.* London: Department of Health.

Department of Health, Chief Medical Officer. 2005. *The Expert Patients Programme* [Online: British Department of Health]. Available at: http://www. dh.gov.uk/en/Aboutus/MinistersandDepartmentLeaders/ChiefMedicalOfficer/ ProgressOnPolicy/ProgressBrowsableDocument/DH_5380844> [accessed 15 January 2009].

Douglas, C.M.W. 2009. *The Social Construction of Patients, their Involvement, and New Medical Technologies: The Case of Pharmacogenetics.* Unpublished PhD Dissertation: University of York.

Epstein, S. 2008. Patient groups and health movements, in *The Handbook of Science and Technology Studies,* edited by E.J. Hackett et al., 3rd Edition, Cambridge, MA: MIT Press, 499-539.

Foucault, M. 1973. *The Birth of the Clinic: An Archaeology of Medical Perception.* London: Routledge.

Hanley, B. et al. 2003. *Involving the Public in NHS, Public Health and Social Care Research: Briefing Notes for Researchers Report to INVOLVE.* Second edition, Eastleigh: Department of Health.

Hedgecoe, A. and Martin, P. 2003. 'The drugs don't work': Expectations and the shaping of pharmacogenetics, *Social Studies of Science,* 33, 327-64.

Home Office. 2006. *Crime Statistics – More Serious Violence for England & Wales Apr. 2005-Mar. 2006.* Available at: http://www.crimestatistics.org.uk/tool/ Default.asp?region=0&force=0&cdrp=0&l1=6&=1&l3=27&sub=0&v=36 [accessed 27 November 2006].

Pirmohamed, M. and Lewis, G. 2004. The implications of pharmacogenetics and pharmacogenomics for drug development and health care, in *Regulating the Cost and Use of Pharmaceuticals in Europe,* edited by E. Mossialos et al. Maidenhead, England: Open University Press, 279-96.

Reed, T. 2006. Fool Britannia – a society in denial. Available at: http://news.uk.msn. com/Fool_Britannia_A_Society_In_Denial.aspx [accessed 23 November 2006].

Webster, A. et al. 2004. Integrating pharmacogenetics into society: In search of a model. *Nature Reviews Genetics,* 5, 7-13.

Webster, A., Douglas, C.M.W. and Lewis, G. 2009. Making sense of medicine: 'Lay pharmacology' and narratives of safety and efficacy. *Science as Culture,* 18(2), 233-347.

Zola, I.K. 1973. Pathways to the doctor – From person to patient. *Social Science & Medicine,* 7, 677-89.

PART 2
Governing Bio-Objects

The second part of the book concentrates on analyses of current governance practices related to new forms of life.

Nik Brown's chapter analyses the ways in which purity of the human is first constructed and then politically separated from the animal in the case of trans-species embryos and their regulatory politics in the UK. Brown claims that the prevalent way of talking about human and privileging human as the constitutive object of politics in UK's parliamentary debates points to the fact that interspecies embryos are not politically treated as new forms of life at all. Instead, they are reduced to particular established discursive orders to make sense of the embryos. The question of newness, then, is reduced to the calculation exercise of animal and human genetic material in the embryo instead of treating this vital matter as a new form or category of life. In this sense the transspecies embryos are, in fact, "stripped" from their new beingness as the newness of human and animal genetic assemblage is denied in the process. The analysis shows how the bio-objectification is, in this sense, incomplete and conservative in its attempt of keeping the category of "human" pure from the contamination of the animal matters in the practices of genetic governance.

Janus Hansen's chapter takes a comparative look at the coexistence of the regulation of genetically modified crops, its political implementation mechanisms and surrounding debates in three European countries: Denmark, the UK and Germany. Hansen claims that there are three dimensions in national politics that have an impact on how the coexistence debates treat new bio-objects, here GM crops, in relation to older forms of life. The first crucial dimension is the whole discursive configuration of GM crops that affects the ways in which they can, and do, become contested. The second dimension of comparison relates to the ways in which expertise on new bio-objects is put together in its different institutionalized forms, and how this is taken as a requirement for making qualified assessments about the GM crops. Lastly, the idea of public engagement differs in each country in its organization and modes of operation to the degree that the whole idea of "public" involved in these debates radically differs between the countries compared. As such, Hansen argues, bio-objects themselves become co-constitutive of the very publics making political decisions about them. This makes EU-level regulation of such new entities very challenging and calls for more systemic comparison of the local implementation processes involved.

In Chapter 7, Aaro Tupasela takes a look at the logics of care in the practical governance of hereditary diseases. Tupasela specifically asks how the ideal of autonomy and choice of knowing about personal risks becomes possible when faced with the inbuilt preventative imperative of genetic research practices and surrounding governance structures. He claims that in the case of hereditary cancer research the two central ideas of biomedical research translate to a particular logics of care that is witnessed in related practice of phone counselling. Here the responsibilities of knowing and caring, and of autonomy and paternalism, are distributed between the patients and the researchers in a way that defies the traditional ways of seeing autonomy as an individual right and the imperative of non-directiveness for researchers attempting to offer preventive care. In this chapter, we see how new bio-objects, such as hereditary mutations indicating a risk of cancer, challenge old ways of governing and caring both those at risk and the ones offering treatments.

Continuing on these lines of questioning contemporary bio-medical practice Nete Schwennesen shows how genetic risk assessment in prenatal screening situations becomes literally a matter of life and death. Her ethnographic study at an ultrasound clinic in Denmark focuses on the ways in which fetal life becomes a living being or not, thus revealing a situation where the status of life as an object of decision is constantly negotiated against the boundaries of human/nonhuman and normal/pathological. New life becomes a matter of concern through particular visualization and calculation practices. Schwennesen shows how prenatal risk figures work as elements in sense-making practices and for inclusion – exclusion decisions about giving birth to, or aborting life. Interestingly, she puts forward a similar argument seen in Tupasela's chapter, namely that the objectified matter of the not-yet-born-life distributes the relations of governance and responsibility in new, unexpected ways. In this case the relations are actively shaped by the fetus conceived at once as a material, social and discursive bio-object, despite its liminal status in the realms of life.

Sakari Tamminen and Niki Vermeulen

Chapter 5

Beasting Biology: Interspecies Politics

Nik Brown

Introduction – beasting the embryo

> We must learn ... to investigate ... the practical and political mystery of
> separation. What is man, if he is always the place – and, at the same time, the
> result – of ceaseless divisions and caesurae? (Agamben 2004: 16)

From any number of angles, both biological and philosophical, there has been
good reason to assume a fundamental destabilisation of the human as the
pre-eminent figure of Western Enlightenment discourse. As an increasingly
unstable techno-scientific bio-object, the human has been subject to far-reaching
'transbiological' interventions. Nevertheless, this has been coupled with a public
cultural politics in which the human continues to operate as a surprisingly
resilient political object. This chapter explores these tensions between the
science of transbiology and a continuing cultural and political attachment to
the human. In particular, I want to reflect upon the continued privileging of the
human as an object category, even in contexts where the very question of what it
is to be human, or animal, are open to fundamental question.

This chapter reworks an earlier paper (Brown 2009) in which I sought to
unravel controversial regulatory changes in the UK to permit the creation of
transpecies embryos in the late 2000s. Much of this debate focussed on the
problems of metric quantification and calculation. That is, participants in the
debate wrestled with the difficulty of trying to calculate whether an experimental
embryonic entity is human, or animal, or both, and in what percentage
proportions? These were, and are, questions of great consequence. They are
instrumental in channelling embryos down different regulatory pathways
depending on whether they are predominantly human or animal. And yet, the
very instability and hybridity of a transpecies embryo undermines and thwarts
these attempts at classification, for regulatory, biological or political purposes.

These category questions surrounding the legal constitutional status of the
human have been central to Agamben's thinking on the politics of life (2004;
1998). History for Agamben (as for others from Gray 2002 to Derrida 2002 in
very different ways) is the ceaseless attempt to resolve the slippage between
bare life (zoê) and qualified or political life (bios). He writes of the queer
'zones of indistinction' between the human and the nonhuman and that '... the
fundamental categorical pair of Western politics is not that of friend/enemy ...

but that of bare life/political existence, zoê/bios, exclusion/inclusion' (1998: 8). Bare life lies on the threshold between raw and political matter, a threshold which is increasingly indistinct and less amenable to purification. As we will see below, transpecies embryos poignantly illustrate biologically and culturally indistinct zones between animals and humans.

The reference to queer is intentional in this context. It gestures towards a body of theory that interrogates dominant normative positions organised around gender, sexuality, race or indeed even species. Hybrid embryos are the uneasy products of a subversive science that both morally clings to, and yet biologically contaminates, human specialness.

What emerges from these debates is, in fact, a highly incalculable bio-object that defies numbering and classification. The chapter attempts to make connections between interspecies and intersex in the work of Kath Weston (2002). Weston writes of the incalculability of intersex and posits the numberless 'zero' as means of theorising forms of sex that defy metric expression. Zero is, I would argue, equally relevant in theorising the incalculable distributions of the human and the animal in transpecies embryos.

Hybrids occupy ephemeral liminal spaces at the edges of easy classification. Initially, the UK Government had proposed that it might be possible to count the DNA of an embryo in determining whether it should be regulated as a human or an animal embryo. Expert witness to the UK Parliament's enquiry were quick to point out the difficulties of this: '... My view would be that the essence of what makes us human as opposed to other animals is not easily measured in DNA ... Somewhere deep down inside there must be specific DNA determinants that make me a person and a colleague whom I will not name a rat...' (Prof. Bobrow, Q936, JCHTE July 2007).

This chapter focuses on the debate in the Westminster parliament and particularly the Joint Committee of the House of Lords and Commons on the *Draft Human Tissue and Embryology Bill* (2007) which sat throughout June 2007. It builds methodologically and analytically on previous analyses of parliamentary discourse on reproductive technology (Mulkay 1997, Franklin 1991, Kirejczyk 1999, Parry 2003).

My concern here is with the management of the mess, dirt and the untidiness of bioscience governance. In taking up these themes I want to pursue the tension between the differentiation and dedifferentiation of interspecies bodies. Whilst the UK government initially sought to accurately distinguish embryos derived from fundamentally different techniques, many from the research community argued that transpecies embryos should not be treated as different to one another in principle. Given that they are all equally hybrid, the argument went, they should all be allowed in law and regulation. This tension between technical difference and moral sameness became one of the central terms of the debate. Before discussing this contradiction in more detail it is worth offering a brief overview of the events which led to the most radical overhaul of reproductive legislation of recent decades.

The UK transpecies embryo debate

The trigger for the most recent debate in the UK came in early 2006 when research teams released details of their intention to seek licences to create transpecies embryos from the UK Human Fertilisation and Embryology Authority (HFEA). The aim was to use non-human eggs or ova combined with nuclear human DNA in order to derive stem cells, giving rise to 'cytoplasmic hybrid embryos' or 'cybrids' having a human nucleus and non-human 'outer egg layer' or 'shell'. The cybrid is only one of at least four or more types of transpecies hybrid debated. The existing legislation, dating to the Human Fertilisation and Embryology Act of 1990, had effectively outlawed the deliberate creation of transpecies human embryos. Although, as discussed below, the original meaning of the Act and whether cybrid embryos were indeed proscribed by it became the focus of debate and scrutiny by the regulator and in the High Court.

By November 2006, the HFEA had received two research applications effectively forcing the hand of policy-making to clarify the regulatory position amidst continuing uncertainty and debate over a whole range of scientific issues including the biological value of stem cells derived from genetically unstable sources; the comparability of human and nonhuman models for stem cell derivation, etc.

The question of hybrid transpecies embryos remained largely unresolved in a British regulatory context where the banning of such embryos had been foundational to the earlier 1990 Act. Prompted by growing research interest the Government moved throughout 2006 to clarify its position. Under the recommendation of the UK Chief Medical Officer, the Government's White Paper (*Review of Human Fertilisation and Embryology Act*) proposed that '… the creation of hybrid and chimera embryos *in vitro*, should not be allowed' (December 2006). This provoked resolute scientific opposition with the Government seen as unduly cautious in response to a small scale public consultation suggesting a potential public backlash. *The Draft Human Tissues and Embryos Bill* (May 2007) was substantially in keeping with the earlier White Paper prompting further organised challenges from the science lobby and was hotly-debated during the proceedings of the Joint Committee of the House of Commons and Lords taking evidence and witness statements throughout June 2007.

Those favouring transpecies research cohered to form a formidable institutional alliance. The licence applications highlighted regulatory uncertainty over the existing powers of the HFEA and the exact legal distinction between several classes of transpecies body. 'True hybrids' – created by fusing two complete gametes of human and non-human origin – were seen to be explicitly forbidden for research by the 1990 legislation. Less clear was whether cytoplasmic or cybrid embryos – created by injecting a human nucleus into a non-human egg – were proscribed in the same way. The question for policy was whether cybrid embryos should be legally viewed as 'hybrid' in which case they were forbidden under existing legislation or 'human' and permitted for research purposes.

Following lengthy deliberation, the regulator determined that cybrids were to be interpreted as essentially human, and therefore wholly acceptable, under the provisions of the earlier legislation. The ruling relied on a view that came to prevail throughout the debate that the mitochondrial DNA of the animal egg was insignificant in determining the species attributes of the resulting embryo. In effect, the nuclear DNA of the cybrid would be taken to legally define the species attributes of the embryo. Heritable mitochondrial egg DNA was to be disregarded. This privileging of the human and the nuclear is discussed in much greater depth below and became central in defending and promoting the legitimacy of research on cytoplasmic hybrids.

Although the HFEA had ruled that the licence applications now fell within its jurisdiction (HFEA 2007a), it decided not to act without first holding a second and more wide-ranging public consultation (April-July 2007). But far from being an 'up-stream' consultation it was conducted alongside and arguably after decisions had been taken and after repeatedly stating its policy commitment to approve cytoplasmic embryos for research. For example, as early as August 2005, the HFEA commented that pure hybrids (created from the equal fusing of human and nonhuman gametes) are 'consistent with research as it is currently allowed' (HFEA 2005). Its April 2007 consultation document expressed a desire to see that 'the current law, which permits the creation of hybrids only for very limited purposes ... be extended so that hybrid embryos can be created for the same research purposes as other embryos' (HFEA 2007b: 13).

The revised Bill (*Human Fertilisation and Embryology Bill*) was published in November 2007. By now the Government had moved entirely from its earlier December 2006 position abandoning the prohibition of true hybrids. The publication of the *Parliamentary Joint Committee Report on the Human Tissues and Embryos Bill* (August 2007) played a crucial role in redefining a more permissive framework for the legislation which received its final hearing in October 2008. In what follows, this chapter focuses on the passage of the transpecies embryo through the narrative and historical corridors of UK parliamentary process, and in particular the discussions and witness testimony presented to the Joint Committee.

The Un/Differentiated Embryo

At the centre of this debate lies fundamental tension. There has been ongoing diversification in the numerous techniques used to produce embryonic entities. These differences in technique have given rise to the equally troublesome question of morally distinguishing between one transpecies hybrid embryo and another. In parallel to this technical heterogenisation we have seen concerted efforts to sort out which practices should be allowed and which should not. The original draft bill (May 2007) distinguished between five different kinds of 'inter-species embryos'. Three of which, were to be regulated as permissible for research. The remaining embryos were either to be banned (as in the case of true human hybrid embryos) or regulated by the Home Office instead (*Animals Scientific Procedures Act 1986*) because they were seen to be more animal than human. For the Government, the Chief Medical Officer, had argued firmly that allowing cybrid embryos whilst banning pure hybrids recognised both technical and moral differences between types of transpecies embryo:

> ... the process involved and the resulting entity is different in character to the first. I am surprised that you cannot see that. I think the outside world would see that distinction... (Liam Donaldson, Q249, JCHTE July 2007).

Now, in opposition to this technical and moral heterogenisation, many in the embryological research community sought to treat inter-species hybrids in morally equivalent terms, regardless of technical differences in their creation. That is, to treat all of these transpecies hybrids symmetrically, and in so doing, resolve the seemingly arbitrary moral distinctions that have resulted in the unbanned banned.

> ...I must say that I am very confused by the list and particularly by the exclusion ... of true hybrids, the fusion of animal and human gametes, which apparently are forbidden for research but are allowed in the hamster test. So I would ask what the moral distinction [is] ... . (Prof. Blakemore, Q35, JCHTE July 2007)

This position was consistently adopted by advocates of permissive regulation, including written statements by institutional scientific bodies. A memorandum by the Academy of Medical Sciences similarly switches between a scientific register that recognises and acknowledges technical differences between transpecies bodies, and a moral register in which transpecies bodies are treated as morally equivalent to one another:

> The reasons for prohibiting the creation and use of 'true' hybrid embryos ... while permitting research involving other types of human embryo incorporating animal material, are not clear to us. While we are not aware of any current scientific reasons to create such entities, we cannot rule out the emergence of valid reasons in the future. (Ev 84, JCHTE July 2007)

Nevertheless, arguing that all transpecies embryos should be treated in equivalent terms was in tension with the ubiquitous claim that cybrid embryos (made using a human nucleus and animal egg) were largely human and therefore more legitimate. It has been common in the debate to invoke statistical percentages in referring to cytoplasmic embryos or cybrids as '99.9 per cent human'.

> All that one can do is try to get the essence of the story over and make clear that, as with the cytoplasmic embryos that they are 99.9 percent human, it is just that mitochondrial DNA is there ... . (Mr Walsh, Q294, JCHTE July 2007)

This line of rhetorical defence has been widely taken up and accepted. Casting the discussion in percentage terms has long been a feature of the debate. Going back to the deliberations in the President's Commission on Bioethics in 1998, following the creation of a bovine-human cytoplasmic hybrid by Advanced Cell Technology Inc, there was intensive debate over what the proper nomenclature for these kinds of entities should be. Whether, for instance it was appropriate to refer to them as hybrids or chimeras or something else. Even at that early stage, the tendency on the part of scientific witnesses is to stress the decisive importance of the nuclear DNA over the mitochondrial DNA of the egg or ova. That relative ratio of 99.9 per cent nuclear and 0.1 per cent mitochondrial DNA was seen to be decisive:

> Dr Meslin: Just so we're clear: what would you refer to it [the cybrid] as?
>
> Dr Brinster: I think it would be a dog, because ... the nucleus of the species transplanted into the oocyte [egg] would actually multiply ... and the proteins would become, for example, the proteins of a dog. And it's likely that mitochondria would be transplanted with the nucleus ... . (*Proceedings of the National Bioethics Advisory Commission*, Nov 1998, day 1)

The contradiction here is that this line of defence conflicts with claims that all transpecies hybrids should be treated in the same way, having no real moral distinguishing features as either human or animal – regardless of proportionality. There is then another tension that arises here between these claims for difference and sameness. On the one hand, cybrids are regarded as different or special by virtue of their proportionally greater humanness (99.9 per cent human). The cybrid is a special case. On the other hand, it is argued that all transpecies hybrids should be regarded as essentially the same because of their interspecies origins:

> Chris Mole: ... do you think there is any distinction between what are called true hybrids and the others...?
>
> Dr Lovell-Badge: Fundamentally, no, I do not; I think they should all be treated the same way. The distinction between them is not great and there may be perfectly good reasons [to make them]... (Q645, JCHTE July 2007)

Numbers have been a routine reference point in the metric calculation of humanness across any number of bioscience areas over the last two decades or more. The percentage of genes or DNA shared, or not, between humans and distant evolutionary relatives was standard fare in news and commentary throughout the course of the Human Genome Project. Numbers are emblematic of an exacting scientific epistemology (Porter 1995, Hacking 1983). Ayrn Martin points to the paramount place of enumeration in science and that '... whether counting species, continents, or colours of the rainbow, ambiguity invites the communal elaboration of rules and techniques for defining and separating entities...' (Martin 2004: 944).

It is highly appropriate for the question of interspecies hybrids that Martin's focus is the historically variable question of human chromosome counting. First put at 48 from the 1920s but then revised down to 46 in the 1950s, the case nicely illustrates the contingencies of counting and the categorical judgements that structure what is and is not a countable category.

In terms of the species question in the Human Genome Project, Holmberg writes of interspecies 'ontological gerrymandering' in the way percentage estimates of species affinity ranged wildly: '... boundaries are drawn between humans and non-humans with the help of numbers, the certainty produced by absolute and percentage figures conceals that the categorisations could be made in other ways' (Holmberg 2005: 24). Now, in the context of the Human Genome Project, a small percentage difference between humans and another species really seemed to matter. It metrically separated humans from other primates, drosophila, nematodes and a host of other species. However, here in the transpecies embryo debate small percentage (mitochondrial) differences between humans and other species is presented as not mattering at all.

Jonathan Marks' *What it Means to be 98% Chimpanzee* (2002) explores just this question of the anthropological relevance of genetics for the ongoing reinterpretation of the evolutionary relationship between humans and other species. He focuses on the cultural and historical arbitrariness of the use of genes and numbers in claims to species affinity and non-affinity. His point is that whilst humans and chimps have an estimated 98 per cent of their genetic material in common (and what this means is always open to question) that does not tell us very much about humans, other primates or indeed genes and DNA. It should not, in particular, be taken to suggest that we are mainly chimpanzee, nor that chimpanzees are mainly human. 98 per cent is a plastic rhetoric and small differences can be disproportionately significant.

Participants in the transpecies embryo debate have faced the unenviable challenge of agreeing the numerical percentage threshold at which an embryo should, or should not, be considered human or transpecies or both. The government had proposed that where the weight of humanness exceeds 50 per cent, that embryo should be considered subject to the newly formulated human reproductive legislation. Were the weight of humanness to fall below that threshold then it would be regulated by the Home Office as a question of animal experimentation. Witnesses to the committee were very sceptical about this:

> Robert Key: has [the Government] has got it right... if there is 50 percent or more that is of human genetic constitution then you should define it as human. Is that distinction scientifically sound?

> Dr Minger: It goes back to ... what is 50 percent? Is it 50 percent DNA, is it 50 percent genes, is it 50 percent coding sequences, is it 50 percent chromosomes? That concept to me again biologically just does not make any sense.... (Q651, JCHTE, July 2007)

There are then acute ambiguities in these discussions of percentages. There is acknowledged uncertainty over whether the Government's 50 per cent proposal is scientifically sensible. Whilst, at the same time, many in the debate continued to use percentage based arguments to defend the creation of cybrids as predominantly human and insignificantly nonhuman. Whilst numbers offer the promise of precision they have proven difficult to stabilise in settling disputes here on the boundaries of humanness. Kath Weston similarly reflects on the metric numbering of gender, and in particularly the less stable ambiguous genders. Intersex, like interspecies, '... unsettles the presumption that ... gender must ultimately refer back to genders – something countable and enumerable. ... whether that sequence extends to two (Man, Woman) or three (Man, Woman, Third Gender)... Yet, as a zero concept, unsexed is no androgyny' (2002: 40-41). Her point is that the numberless zero, like interspecies and intersex bodies, is often inappropriately conceptualised through the numbered binaries of male or female, human and nonhuman.

These problems with the numbering of difference are compounded by time and process. The hybridity of a transpecies embryo is far from static but changes over time and throughout the sometimes short gestational development of transpecies embryos. What may fall under one definitional umbrella for the purposes of licensing and law may well develop into something else in this far from exacting area of developmental biology. It may become possible for these kinds of entities to morph from one kind of hybrid to that of another during early gestation – and in so doing, migrate from one legal definition to that of another.

> Dr Lovell-Badge: it is very hard to come up with any strict definition saying this is 50 percent human and 50 percent animal, therefore it falls into this category rather than this one, because things change ... you may start off with an embryo which is 20 percent human and end up with something which is 60 ... or vice versa (Q621, JCHTE, July 2007).

> Dr Gibson: Given all the interactions between nuclear mitochondrial DNA, a lot of which you do not know about yet, it opens up ... the possibility of strange happenings.

> Prof. Holm: ...as has been hypothesised ... the animal mitochondria will eventually disappear and be replaced by human mitochondria, in which case they would at some point undoubtedly, on anyone's definition, be fully human.... (Q654 JCHTE, July 2007)

This question of the relationship between the nuclear and mitochondrial DNA, though shrouded in uncertainty, was far from uncontested. A member of one regulatory body and a specialist in mitochondrial genetic diseases contested the claim that nuclear DNA determines the genetic makeup of the cell (personal communication). Ultimately the claim is inconsistent with the incidence of hereditary mitochondrial disease. Whilst rare these are well characterised and inherited exclusively through the matrilineal line and passed on through mitochondrial egg DNA. The member was rather perplexed that the matrilineal female mitochondrion of the animal egg was being downplayed in this way.

This touches on what science studies scholarship has written about as the gendering of the cell (Haraway 1995, Hird 2002a, 2002b, M'charek 2005). Over the course of the debate there were searching questions over whether, or not, the human nuclear DNA would ultimately determine the species attributes of the cybrid or cytoplasmic hybrid, and specifically the egg/ova mitochondrial DNA. In other words, any embryos created would become more human over time. The nuclear DNA would come to dominate the enucleated egg, overwhelming the nonhuman mitochondrial DNA of the ova. The debate adds a speciesist twist to observations by feminist scholars and historians of science about the hierarchical positioning of the 'male' nucleus over the female mitochondrial periphery. In the debate we have a set of dichotomies that pitch the human nuclear against the subservient nonhuman mitochondrial egg DNA.

Nuclear/Mitochondria
Male/Female
Human/Animal
Speciated core/Despeciated shell
Timeless/Finite
Permanent/Impermanent

Alongside this, the debate has also witnessed some significant shifts in rhetoric. Stephen Minger, one of the key licence applicants to undertake work on transpecies embryos switched from using the term cell 'enuclation' to 'despeciation' when describing the process by which nonhuman DNA is removed from egg cells. Just as importantly, during the course of the debate the UK Department of Health substituted the term 'transpecies' embryos with 'human admixed embryos' in an effort to emphasise the human over the nonhuman.

These questions of hybrid identity have important implications for regulatory oversight, especially in terms of species boundaries between regulatory authorities. The problem with a 'true hybrid' – created by the equal fusing of nonhuman and human gametes – is that it completely confuses the institutional margins of human and nonhuman regulatory authorities, between the HFEA and the Home Office respectively. This has been written of elsewhere as a question of institutional species identity (Brown et al. 2006). Regulatory bodies are varyingly modelled on assumptions about the species characteristics of regulated bodies – animals and humans. Today, many of the institutional species assumptions that demarcate regulatory agencies across the human and the nonhuman are being disrupted in cases like this. Lobbyists for permissive legislation had argued for a simple overarching definition of the embryo and against the Government's efforts to differentiate techniques and resulting embryos. This however came into conflict with the species structure of regulation.

> Prof. Bobrow: The idea of a single, general definition is very attractive. The problem for me in this area is that we do have this bipid system where we regulate animals and things that come from an animal in one way and things that start as human in another way. We are trying to deal with the meeting of it (Q943, JCHTE, July 2007)

Another problem surfaced in relation to the mandate of the Home Office which has no responsibility for any embryos until half way through their gestational development. It is then conceivable that a hybrid animal-human embryo may go completely unregulated until the midpoint in its development. Whilst this feature of law could result in under-regulation it could also conceivably result in over-regulation where gestation may be very short, as is the case with mice embryos which reach the gestational mid-point at about nine and a half days. As one witness put it, an interspecies mouse may be subject to greater protection than a human embryo because of '… the administrative inconvenience of having to coordinate two regulatory regimes which is not to me a very persuasive argument' (Q940, JCHTE, July 2007). There is then a conflict between the institutionalisation of hybrid regulation and the more fluid aspects of scientific innovation and biology. In other words, the interface and borders between regulatory bodies are likely to be of greater importance than the individual regulatory characteristics of agencies.

What can be seen throughout these discussions is that science's more traditional epistemological commitment to differentiation, taxonomy and distinction has been set aside. Most scientists consulted in the debate argued strongly that whilst technically different, the various classes of transpecies embryo should be regarded as morally and culturally equal. Nevertheless, this moral dedifferentiation is at odds with the presentation of the cybrid embryo as proportionally more human, and therefore more legitimate for human reproductive research. The marginalisation of the nonhuman mitochondria, drawing on a historical gendering of the cell, was strategically important in this respect. The attempt to flatten the moral terrain of reproductive research ultimately undermined the Government's argument that embryos produced by different means and of different species origins are likely to be publicly received in culturally uneven ways. That moral levelling of embryo types also disrupted Government efforts to unambiguously demarcate the regulatory zones of animal and human research.

Conclusion – Beasting the embryo

In November 2008, the UK parliament voted in favour of a radically revised Bill in which the prohibition of 'true hybrids' had been entirely abandoned by the Government. The revised definitions were now much more flexible than originally proposed with the addition of a new and contentious provision for the *ad hoc* regulation of yet unforeseen kinds of transpecies embryonic entity. The final wording of the 2008 Act defines an interspecies embryo as, amongst other things, '... any embryo... in which the animal DNA is not predominant' (4, 6, e). An earlier version of this section had however been heavily criticised for its ambiguity and for failing to adequately distinguish between the differing regulatory responsibilities of the Home Office (research animals) and the HFEA: 'For the purpose of this Act an inter-species embryo is ... *such other thing* as may be specified in regulations' ([my italics] Draft Bill, clause 4.2). It was argued that both formulations ('any embryo' and 'such other thing') would 'future proof' the legislation against painstaking revision and enshrine the novel unpredictability of hybrid biology in legal statute. It would give the Secretary of State for Health the powers to add new inter-species entities as and when they arise. Nevertheless, the meaning of 'predominant' in the final Bill and its stability both biologically and legally remains unclear. The section represents a curious compromise between different sides in the debate and is a concession that indirectly preserves the suggestion of a 50 per cent division in settling the regulatory identity of interspecies embryos. One the one hand, the clause biologically acknowledges the instabilities of genetic species identity, but then institutionally persists with an increasingly troubled partitioning of human and animal governance. Hybrids are profane and unruly in this way signalling the frailty of the 'modern constitutional' order and threatening to overwhelm legislative boundaries that are arguably already outmoded even before they become law (Latour 1993).

The story told here traces a debate in which an organised lobby succeeded in steering and directing the political reproductive agenda. To this extent, hard classificatory borders have been softened and the traditional taxonomical regime of science (as a 'difference machine') has been substituted with an open-ended moral pragmatism. Rigid regulatory orders have been replaced with suitably leaky ones, advancing the interspecies direction of an emerging embryological research order.

Important in explaining the outcome of this debate has been the moral flattening and homogenisation of the different classes of embryos described in the legislation. The early attempts by the UK Government and its chief medical officer to acknowledge in law potentially varied cultural and political readings of, for example, 'pure hybrids' as distinct from 'cybrids', had failed. Advocates of 'permissive' legislation from the embryological research community had successfully petitioned for a dedifferentiation of the moral and cultural landscape that applied to research embryos. Difference would be levelled within what Agamben (1998) calls a generalised 'zone of indistinction' smoothing an otherwise technically and morally uneven area of scientific research. Indistinction implies just this kind of erasure in which the unevenness of embodied difference and species identity is reduced to the bareness of life itself.

The hybrid has become the central expression of contemporary corporeal and institutional innovation. Hybrid bio-objects (both biologically and socially) are, in cases like this, embroiled in an ongoing process of 'institutional biosociality' (Brown and Michael 2004) and mutual revision. The making of regulatory zones like the UK's reproductive space is the outcome of an on-going and unfinished reclassification of innovated nature and political jurisdiction. It remains to be seen whether and to what extent these kinds of corporal and institutional fluidities are practically worked out in the course of regulatory business. The vacuum that lies between the duties and responsibilities of animal (Home Office) and human (HFEA) regulatory bodies is likely to be a continuing source of controversy. The legacy of existing regulatory order remains problematic for the governance of simultaneously transpecies and transinstiutional hybrid bodies. Returning to Weston's discussion of intersex, her purpose has been to elucidate those occasions where the numbering of gender is disrupted or 'zeroed'. Intersex and interspecies bodies trouble the metric rationality of taxonomical order and challenge the normative boundaries separating the human from the nonhuman. Male and female, human and animal defy precision here at the margins of institutional and embodied hybridity. The contradictory instabilities of the zero, the numberless, are very unlikely to be resolved by legislation that preserves a binary separation between the regulation of the human and the nonhuman. The contrasting poles of the human and the nonhuman remain precarious and unstable havens into which regulatory policy-making retreats when confronted with the monstrous.

Crucially, it is important not to mistake the creation of interspecies bodies for an erosion of the humanist species hierarchies to which Agamben refers in the opening passage of this paper. The sustained effort to stress, for example, the humanness of cytoplasmic embryos illustrates the continuing privileging of the human and not a cultural erosion of species boundaries. Again echoing Latour's modern constitution above, the notion of 'beasting' applies asymmetrically. That is, the gathering pace of hybrid interspecies research is radically at odds with a cultural emphasis on the humanness of beasted embryos. It is certainly the case that this debate has involved a far reaching transformation of species boundaries in terms of paving the way for an increased range of embryological transpecies research. Nevertheless, the recent legislation represents a constitutional refortification of traditional speciesist hierarchies.

References

Agamben, G. 1998. *Homo Sacer: Sovereign Power and Bare Life*. Stanford, CA: Stanford University Press.

Agamben, G. 2004. *The Open: Man and Animal*. Stanford, CA: Stanford University Press.

Brown, N. 2009. Beasting the embryo: The metrics of humanness in the transpecies embryo debate. *Biosocieties*, 4, 147-163.

Brown, N., Faulkner, A., Kent, J. and Michael, M. 2006. Regulating Hybrids – 'making a mess' and 'cleaning up' in Tissue Engineering and Xenotransplantation. *Social Theory and Health*, 4, 1-24.

Brown N. and Michael, M. 2004. Risky Creatures: Institutional species boundary change in biotechnology regulation. *Health, Risk and Society*, 6, 207-22.

Department of Health 1990. *Human Fertilisation and Embryology Act*. London: Stationery Office.

Department of Health 2006. *Review of Human Fertilisation and Embryology Act*. London: Stationery Office.

Department of Health 2007. *Draft Bill: Human Tissues and Embryos*. London: Stationery Office.

Department of Health 2008. *Human Fertilisation and Embryology Act*. London: Stationery Office.

Derrida, J. 2002. The animal that therefore I am. *Critical Inquiry*, 28(2), 369-418. European Patent Office 1999, EP 380646.

Franklin, S. 1999. Making representations: The parliamentary debate on the Human Fertilisation and Embryology Act, in *Technologies of Procreation: Kinship in the Age of Assisted Conception*, edited by J. Edwards et al. London: Routledge.

Gray, J. 2002. *Straw Dogs*. London: Granta Books.

Hacking, I. 1983. *Representing and Intervening*. Cambridge: Cambridge University Press.

Haraway, D. 1995. Otherworldly Conversations, Terran Topics, Local Terms, in, *Biopolitics: A Feminist and Ecological Reader on Biotechnology*, edited by V. Shiva and I. Moser. London. Zed Books.

Haraway, D. 1997. *Modest_Witness@Second_Millennium. FemaleMan Meets_ OncoMouse: Feminism and Technoscience*. London: Routledge.

Hird, M.J. 2002a. Re(pro)ducing sexual difference. *Parallax*, 8(4), 94-107.

Hird, M.J. 2002b. The corporeal generosity of maternity. *Body and Society*, 13(1), 1-20.

Holmberg, T. 2005. Questioning the number of the beast: Constructions of humanness in a Human Genome Project. *Science as Culture*, 14(1), 23-37.

House of Commons Science and Technology Committee 2005. *Human Reproductive Technologies and the Law*, Vol. I-II. London: Stationery Office.

House of Lords, House of Commons 2007. *Joint Committee of Human Tissue and Embryos (Draft) Bill* (Aug), Vol. I-II. London: Stationery Office.

House of Lords, House of Commons July 2007. *Joint Committee on the Human Tissue and Embryos (Draft) Bill, Minutes of Evidence*. London: Stationery Office.

Human Fertilisation and Embryology Authority 2005. *Response by the Human Fertilisation and Embryology Authority to the Department of Health's Consultation on the Review of the Human Fertilisation and Embryology Act. Report number 05/33273*. London: Crown Copyright.

Human Fertilisation and Embryology Authority 2007a. *HFEA Statement on its Decision Regarding Hybrid Embryos*, 5 September. Available at: www.hfea. gov.uk/en/1581.html [accessed November 2008].

Human Fertilisation and Embryology Authority 2007b. *Hybrids and Chimeras – A Report on the Findings of the Consultation*. London: Crown Copyright.

Kirejczyk, M. 1999. Parliamentary cultures and human embryos: The Dutch and British debates compared. *Social Studies of Science*, 29(6), 889-912.

Latour, B. 1993. *We Have Never Been Modern*. Cambridge, MA: Harvard University Press.

Marks, J. 2002. *What it Means to be 98% Chimpanzee*. Los Angeles, CA: University of California Press.

Martin, A. 2004. Can't anybody count? Counting as an epistemic theme in the history of human chromosomes. *Social Studies of Science*, 34(6), 923-48.

M'charek, A. 2005. The mitochondrial eve of modern genetics: Of peoples and genomes, or the routinization of race. *Science as Culture*, 14(2), 161-83.

Mulkay, M. 1997. *The Embryo Research Debate: Science and the Politics of Reproduction*. Cambridge: Cambridge University Press.

Parry, S. 2003. The politics of cloning: Mapping the rhetorical convergence of embryos and stem cells in parliamentary debates. *New Genetics and Society*, 22, 145-68.

Porter, T.M. 1995. *Trust in Numbers: The Pursuit of Objectivity in Science and Public Life*. Princeton, NJ: Princeton University Press.

Sheng, H. et al. 2003. Embryonic stem cells generated by nuclear transfer of human somatic nuclei into rabbit oocytes. *Cell Research*, 13, 251-63.

Weston, K. 2002. *Gender in Real Time: Power and Transience in a Visual Age*. New York and London: Routledge.

Chapter 6

Comparing Public Engagement with Bio-objects: Implementing Co-existence Regimes for GM Crops in Denmark, the UK and Germany

Janus Hansen

Some of the first artificially produced, novel forms of life to migrate from laboratories into commercial circulation were genetically modified organisms (GMOs) to be used in agriculture. From early on scientists, industrialists and policy makers invested high hopes and significant sums of money in the expectation that GMOs would improve agricultural productivity and provide lucrative returns on investments in research and development. However, this technological trajectory has proven highly controversial, and GM crops have turned into one of the most contested applications of molecular biology in Europe (Torgersen et al. 2002).

The struggles to stabilise GMOs as bio-objects of governance have produced significant disruptions in the interaction between science and society. As a consequence, the governance of biotechnology today needs to balance the desire to stimulate research and innovation with efforts to ensure public acceptance and legitimacy of innovation much more actively than before. GMOs have served as a catalyst for a number of public concerns in relation to the rapid developments in modern biology, such as risk management, long-term socio-economic consequences of biotech innovation as well as (bio)ethical issues. Issues related to the governance of GMOs therefore provide a good place to examine some of the challenges posed by the general movement towards an increased public engagement with bio-objects as a means to ensure their public acceptability and legitimacy.

In STS and adjacent academic fields an understanding has been nurtured that when (members of) 'the public' express concern about new technological developments, it is desirable that the public is consulted in some way or another (Renn et al. 1995). However, much of the research interest in participatory governance is manifested either in rather abstract and general terms – often advocating normative principles ideally to be fulfilled in participatory processes (e.g. Sclove 1995, Fischer 2000, Durant 1999) – or in single case studies or evaluations of particular specific engagement events (e.g. Marris and Joly 1999, Horlick-Jones et al. 2007). More systematic explorations of the interaction between ingrown institutional routines and politico-cultural practices, on the one hand, and public engagement with novel kinds of bio-objects, on the other, are scarcer (see Joss and Bellucci 2002, Hansen 2010). As Gottweis (2008) argues there is no such thing as a 'pure' public to engage. Therefore, it is important to examine more thoroughly how bio-objects interact with 'culturally derived normativities' (Gottweis and Petersen 2009). In this chapter, I review how the controversies over GMOs have created different scopes for public engagement in different political settings. More specifically, I examine and compare the implementation of coexistence regimes for GMO and non-GMO crops in three EU countries with particular attention to the possibilities created for public engagement.

Exploring variety in the legitimatory practices on bio-objects

The implementation of regulatory regimes to ensure the coexistence of GM crops with conventional and organic agriculture provides a good case for exploring how public engagement with bio-objects is embedded in politico-cultural settings, which may imbue such objects with different social meanings. The legislation is the same across the EU, the implementation took place more or less simultaneously and it refers largely to the same scientific knowledge base. This means that some of the spatio-temporal fluidity characteristic of bio-objects (see Introduction) can be managed for the purpose of comparative research. In many EU countries the implementation of coexistence regimes was accompanied by some kind of public consultation as a means to enhance the legitimacy of the regulatory frameworks. Yet, these processes exhibit some notable differences, as I shall elaborate in the following.

The aim is to compare similarities and differences in the institutional settings and the processes through which regulatory schemes were produced. This will show how the introduction of commodified bio-objects is accompanied by rather different opportunities for public engagement, which in turn shapes the bio-objectification processes. For this purpose, I briefly recount how coexistence regimes were established in Denmark, the UK and Germany. These countries are selected because they have all experienced profound public controversies over GM crops, and they have all taken initiatives to engage 'the public' in reaction to those controversies – though in different manners, giving different interpretations as to what constitutes proper public engagement. Subsequently, I contrast these cases with regard to three issues, which designate important scope conditions for public engagement, namely:

1. The discursive configuration of the lines of contention in the respective policy arenas.
2. The organisation of scientific expertise for policy making.
3. The ensuing channels for inputs from 'the public' into policy formation.

The comparison builds on an interpretive analysis of policy documents, media sources and stakeholder communication (much of which is reported in more detail elsewhere, [Hansen 2010]). This short chapter obviously cannot account for the full complexity of the cases. However, it will serve to illustrate why the politico-cultural context must be taken into account when the complex role of public engagement with bio-objects such as GMOs is analysed. The comparison is deliberately relational. Methodologically, this means that the cases are interpreted through comparison with each other, rather than against any pre-given normative standards of what constitutes 'proper' public engagement practices.

Coexistence: The repoliticisation of a technical detail

Following continued scientific uncertainty regarding the safety of GMOs combined with the outspoken public opposition in several countries, a number of EU member states issued a *de facto* moratorium on the approval procedures for new GM crops and products in 1999. During the moratorium a new regulatory regime was negotiated, which was more restrictive in its demands for scientific risk assessments and foresaw – at least nominally – a larger scope for public consultation.

At this point, only one issue was still outstanding before commercial growth of GM crops could be resumed; the regulation of the co-existence of GMOs with conventional and organic farming (2003/556/EC). Whereas most regulation was put into force for the whole Union uniformly, the regulation of coexistence was left to the individual member states in order to take account of the very different agricultural conditions of cultivation across the Union (2003/556/EC, §7).

'Coexistence' pertains in this context to the physical separation of 'old' and 'new' bio-objects, which should not be allowed to contaminate each other. However, it also designates the search for a (peaceful?) parallel existence of different visions of agricultural production – roughly speaking between high-intensive, industrial agriculture versus organic farming (Levidow and Boschert 2008). Formally coexistence is concerned *solely* with *economic effects* of GM agriculture, as risks to health and the environment allegedly are already dealt with in the Community-wide approval procedures. However, unsurprisingly, this exclusive focus on the economics of GM farming was oftentimes transgressed in public debates, where the full symbolic imagery connected to GMOs repeatedly was re-invoked in the ensuing debates.

At the very top of the elaborate list of 'Principles of the development of coexistence strategies' the Commission places *'transparency and involvement'*. The Commission recommends that "National strategies and best practices for co-existence should be developed in cooperation with all relevant stakeholders and in a transparent manner" (2003/556/EC, Annex § 2.1.1). Although it is by no means given what 'all relevant stakeholders', 'a transparent manner' and 'adequate information' mean in practice, this illustrates the keen awareness of the Commission of the cultural sensitivity affiliated with these bio-objects. It also illustrates the trenchancy of the idea that public involvement can potentially mitigate legitimacy deficits even on issues that are considered essentially technical. As shall be recounted in the following, the implementation of coexistence regimes turned out to be anything but trivial and provided the grounds for yet another round of politicisation of the GMO issue. While many of the parameters of this regulation were given through the common EU framework, the process of arriving at the specific regulatory regimes followed upon and entailed different local interpretations of the need for public engagement, giving different meanings to the need for information provision, transparency and stakeholder involvement. In the following, I account broadly for the central experiences with public engagement with GMOs and more specifically for the formulation of coexistence policies in the three countries. Based on this, I contrast selected aspects of the policy process across the cases.

Denmark

Denmark is usually considered as one of the pioneers in public engagement with controversial technologies. Yet, despite (or perhaps because of) widespread public debate, informal as well as more formalised, on biotech during the 1990s, the population in Denmark has remained one of the most sceptical in Europe in regard to agricultural biotechnology (Jelsøe et al. 2001). This has been reflected in the Danish position in the negotiations in the EU about the general regulatory framework as well as in specific approval procedures. Denmark was one of six countries to introduce the moratorium in 1999 and it was hostile to its removal to the very end (Toft 2007). Since the moratorium was lifted, Denmark has represented one of the most cautious positions in the approval negotiations at the EU level. As such, the governance of GMOs has been comparatively well aligned with the concerns of the wider public – as far as possible within the common EU framework.

The issue of coexistence has been part of the public debate about GMOs in Denmark from early on. It was raised in public as early as 1999 in a consensus conference, where a rather sceptical citizens' panel recommended that producers of GM crops should be taxed to fund a compensatory scheme for potential damages related to GM agriculture; a regulatory tool that has been taken up in several countries, but often fuelling rather than damping down controversies.

Compensatory schemes are controversial not only because they fundamentally permit GM agriculture, but also because the details of the financing and liability rules almost inevitably place competitive disadvantage on either GM or non-GM agriculture, for instance as a levy on GM seeds or as a tax on all farmers (or on public funds). Therefore, what is often presented as a neutral economic instrument inevitably embodies *some* principles of justice and preference *vis-à-vis* GM farming, which is likely to be contested.

The specific procedure establishing the Danish coexistence rules was initiated already while the EU moratorium was in force. In 2002, the Danish Parliament instructed the Government to ensure that GM crops would not spread into the surrounding environment. A group of experts from public research organisations was commissioned to suggest appropriate measures to operationalise this policy goal in practical terms (Vintersborg and Pedersen 2007).

Though essentially technical, the work of this committee dealt with something that was known to be highly controversial. Therefore, the committee was followed by a 'contact group' consisting of a circle of stakeholders representing both supportive and more sceptical stances towards GM agriculture. This included the food industry and the farmers' association, but also the organic farmers' organisation, the Danish Consumer Association and a moderate conservationist organisation (Danmarks Naturfredningsforening), but not the more radical GM opponents represented in Denmark primarily by Greenpeace and Friends of the Earth. The stakeholders did not have any kind of veto power regarding the recommendations, but this dialogue with stakeholders has been stressed repeatedly by the minister for agriculture as a central element in establishing a regime with broad support (e.g. Hansen 2006a). This committee suggested a regulatory regime with four main elements: 1) species-specific separation distances for non-GM and GM varieties, 2) a mandatory training programme for farmers; 3) 'neighbour-dialogue' and notification in a public register for monitoring purposes, and 4) a compensatory scheme in case adventitious contamination should occur despite 'good farming practices' being maintained (Udredningsgruppen 2003). Following upon this expert recommendation, the Danish Board of Technology (DBT) organised a public hearing at the Danish parliament where experts, stakeholders and parliamentarians debated the recommendations (BIOSAM 2004). While there were no wider public engagement initiatives in relation to the formulation of co-existence policies, organised stakeholders expected to represent public interests were consulted from the outset.

Shortly after the public hearing, the Parliament passed a law on coexistence, which was followed by two more operational directives the following year; one regarding cultivation, one on compensation, which was the more controversial element. In the Danish case, it was decided that the compensation scheme should be funded by a levy on GM seeds. In the hearing phase, the GM protagonists claimed this was discriminating against GM agriculture, but expressed satisfaction with the legal clarification, which was expected to facilitate the introduction of GM crops. In the media this framework was presented as paving the way for GM agriculture in Denmark, but the press coverage was generally kept in a neutral tone and only few critical voices were heard. Yet, despite the legal mandate in place to date (December 2010), no GMOs have been cultivated on a commercial basis in open fields in Denmark.

The UK

The politics of GMOs in Britain is heavily influenced by the Bovine Spongiform Encephalopathy (BSE) debacle in the mid-1990s. Although the British government has been one of the keenest proponents of GM technology in Europe, the BSE failure has clearly cast its shadow onto the GMO regulation. Therefore, anti-GM activists, media pressure and a general public uneasy with 'unnatural' food forced the political system to adapt a more restrictive attitude on GMOs than it would have preferred. Following the BSE trouble, the British government had declared a 'new mood for dialogue' (HoL 2000), which created new channels of public influence on the governance of science and technology (Irwin 2006).

One highly significant institutional response to the public unease was the formation in 2001 of the Agricultural and Environmental Biotechnology Commission (AEBC), a strategic advisory body composed of a quite diverse membership in terms of knowledge and attitudes towards GM agriculture. Although the AEBC itself was established to function as a locus for the articulation of public concerns, one of its first recommendations was to instigate a society-wide debate on the prospect of GM agriculture in the UK, which ran in the spring and summer of 2003. The outcome of this 'GM Nation?' debate was a strikingly clear expression of public distaste for GM agriculture. Although the merits of this consultation have been heavily debated and contested (e.g. Horlick-Jones et al. 2007), the British Government was forced onto the defensive in regard to its initial strategy for introducing GM agriculture in the UK. "We had a bad consultation on GM and it set research back in the UK a very long way indeed" as a government official was quoted as saying (*The Guardian*, 17 September 2007). Nevertheless, the British Government continued to work for the introduction of GM agriculture, and the Ministry responsible for agriculture (DEFRA) was given the task of developing coexistence regulation in what must be consider a highly charged and politicised climate.

The first major policy initiative related specifically to coexistence was a report issued by the AEBC in November 2003, which made a number of recommendations regarding the policy principles to guide coexistence. Central among these were the suggestion that the overall policy aim should be regulatory neutrality between different types of crops (AEBC 2003). Most of these recommendations were adapted in a subsequent policy position statement delivered by DEFRA in March 2004 (Beckett 2004). The statement suggested that a code of practice for GM farming should have statutory backing and be accompanied by a compensation scheme for adventitious contamination of non-GM crops. It was stressed that this compensation scheme should be funded by the GM sector itself (ibid.), but no details were given. At the time, it was thought that GM farming could commence shortly thereafter. Yet, it took another two years before DEFRA had formulated a draft for the actual regulatory framework.

In July 2006, DEFRA published a consultation paper, and comments were requested from approximately 130 stakeholder organisations as well as the general public. In response, more than 11,000 comments were received. Many of these were petitions against GM agriculture and pre-printed stock-letters drawn up by organisations campaigning against GMOs, raising a broad spectrum of concerns. Rather unsurprisingly given the public climate, "(t)he responses were polarised between those from groups and individuals who take an essential negative view of GM crops ... and those who do not share these concerns and acknowledge the possibility the GM crops might be grown safely and beneficially" (DEFRA 2007: 2).

It thus proved difficult to accommodate the different interests and viewpoints on coexistence. At least at the time of writing, DEFRA still has not finalised the regulatory framework, and to date no GMOs have been grown commercially in open field in the UK. In the latest public announcement on the topic, DEFRA expects that this will not happen for 'several years' (DEFRA Press Release, 8 November 2007).

Germany

Germany has experienced a longstanding and very polarised debate about agricultural biotechnology, which has migrated into the core of the political system. The environmental movement has achieved a stronger political platform than in most other European countries through the parliamentarian representation and participation in government (1998-2005) of the Green Party.

Since the early 1990s there have been several attempts to mitigate the controversies by various kinds of participatory and deliberative procedures in Germany, though of a more corporatist nature than in Denmark and the UK. For instance the Federal Ministry for Consumer Protection, Nutrition and Agriculture organised a round-table discussion between stakeholder representatives from different sectors of society in 2001-2002. This deliberative procedure was organised at a time when food safety and consumers' ability to choose non-GM products were given a high priority after BSE had been discovered in Germany. Scientists and industry representatives felt that the technology was exposed to irrational fears, misunderstandings and political hostility in the process. As a consequence, the roundtable discussion brought little movement on the landscape of opinion among organised stakeholders, but instead seems to have brought the various stakeholders on either side of the conflict closer together (Hansen 2006b, Hansen 2010).

During 2004 the law regulating agricultural biotechnology in Germany was renewed in order to implement the revised EU directives. The revision was highly contentious both within the German federal political system and among the stakeholders, who were divided in a very polarised fashion. The resulting legislation passed by the Red/Green government was received rather critically by the proponents of GM agriculture, who argued that German legislation practically made the growing of GM crops all but impossible for both research and commercial purposes (Union 2004).

This changed in 2005 when a Christian Democratic/Social Democratic coalition came to power. The incoming government stated its intention to initiate more innovation-friendly policies, and a revision of the law on genetic engineering was announced. After a period of wrestling within the coalition a principled agreement on the need for revisions, including the preparation of a coexistence regime, was set out in February 2007. In this process the Government held bilateral consultations with stakeholders, none of which were carried out in public though.

Judging from the responses of the various stakeholders to this agreement, the political winds had indeed changed when the Green Party was ousted from government. The industry organisations initially expressed their satisfaction with the policy initiative, lauded its 'pragmatic approach' and demanded its immediate implementation in legislation (BDP 2007, DBV 2007). Critics of GM agriculture expressed their outspoken dismay for what they considered a "betrayal of consumers' interests and organic farming" (BÖLW 2007, Verbraucherzentrale 2007).

Yet, during the implementation of the political agreement into actual legislation the GM sceptics managed to mobilise public opinion and gain renewed political leverage. Subsequently, disagreement broke out internally within the CDU/CSU fraction of the Christian Democratic party, with the Bavarian CSU minister of agriculture assuming a more GM critical stance, blocking the approval of specific crops in what seemed to be open violation of EU rules (*Süddeutsche* 2009). The resulting regulatory framework for commercial cultivation identified species-specific separation distances, a duty to notify neighbours prior to sowing, a public register (though no longer open to the general public at a detailed level, to avoid repetition of field destructions by activists) and a legal regime in which farmers can be held liable for economic losses due to contamination of neighbouring fields even if 'good farming practices' are followed. The industry complained that this creates unacceptable legal uncertainties, but rather than opting for a compensatory fund financed by a levy on GM seeds as in Denmark, seed companies accepted to bear potential losses for farmers until the insurance industry is willing to sign policies on liability (biosicherheit.de 2007). During 2008 commercial growth of GM maize has taken off on a limited scale, but farmers have been reluctant to embrace other GM crops.

Comparing regimes of coexistence governance

In all of the three cases the introduction of novel bio-objects in agricultural production has upset the way science, politics and the public conventionally have interacted. The cases share a general trend towards expanded public engagement in the governance of novel technologies. All three cases illustrate that under contemporary conditions, a regulatory problem such as coexistence cannot be dealt with in a purely technocratic fashion outside the public realm without risking that the legitimacy and acceptance of innovation is undermined. Despite the highly technical nature of the issues, the regulatory process is therefore politicised to different degrees in all three countries, and the question of who should count as relevant stakeholders is given a more inclusive interpretation than in routine governance practice. Yet, when looking closer, there are some unique features that distinguish the specific configurations of actors in each of the cases, which means public concerns are manifested in different manners and play different roles in policy making, as will be elaborated in the following.

The discursive lines of contention

The scope for public engagement with the governance of bio-objects is shaped among other things by the way controversies are configured discursively; what is contested and who gets to articulate the concerns of the different positions?

In the Danish case the primary line of contention emerged between the Government, which saw itself obliged to implement EU regulation, on the one hand, and environmental and consumer organisations, on the other. The debate about regulatory principles mostly remained in this relatively circumscribed arena with few actors involved and communication playing out in relatively closed circles without much public attention. The Danish government maintained its relatively reluctant position on GM farming and underlined in public statements that non-GM agriculture, in particular organic farming, must be safeguarded and compensated in case of contamination. As such, the policy goals of the proposed coexistence regime were reasonably well aligned with the still rather sceptical public opinion as far as possible within the demands of EU legislation. Therefore, the organised GM-sceptics focused on the technical details of the regulatory framework (the means), seeking to raise doubts about whether the policy goals could in fact be reached, whereas the ends themselves seemed less suitable for contestation at this point in time.[1] Consequently, there was no immediate potential for a more extensive public mobilisation and broader engagement – at least none appeared.

Compared to this situation, the British case exhibits a bigger and much more visible gap between the agenda of the political establishment and the public opinion. This is seen both in a stronger engagement by the NGO community in the consultation phase, and by a stronger inclination for ordinary citizens to be mobilised in various ways than in both Denmark and Germany. Hence, in the UK a much more direct discursive confrontation between the political establishment and the more general public orchestrated by a host of environmental and consumer organisations can be observed. The goals of the announced policy are contested, rather than simply the means (e.g. Genewatch 2006, DEFRA 2007). In principle, this heightened level of public contestation makes the need for public consultation more pressing, but it also produced a more confrontational climate.

In Germany the major lines of contention are more pronounced within the political system itself, as the opponents of GM agriculture have a stronger foothold within most of the political parties. The discursive controversies involved in the

---

1   For instance, Greenpeace Denmark delivered a technically detailed commentary on the bill during the hearing phase, but did not undertake any activities directed at public mobilisation, which is otherwise one of their preferred modes of operation (Greenpeace DK 2004).

revisions of the GM legislation in general, and the coexistence issue in particular, are therefore played out to a larger extent within the political system itself, rather than as a confrontation between the political system and the wider public. Organised stakeholders thus seek influence through appeals to different fractions of the political system by emphasising, respectively, scientific uncertainties and public hesitation (e.g. Zukunftstiftung 2004) and German competitiveness (e.g. Union 2004). The fact that the political establishment itself was internally divided had the effect that even when the government changed and an allegedly more industry friendly coalition assumed power, the sceptics retained significant political leverage, and the policy revisions turn out to be less radical than the GM protagonists had envisioned (Biosicherheit 2006).

Organising scientific expertise for policy

Public engagement with new technologies is usually seen as a corrective to technocratic dominance of governance (Sclove 1995, Fischer 2000). Therefore, a central challenge when addressing public concerns through extended public engagement is to balance the need for scientific competence and democratic fairness (Renn et al. 1995). Although the relationship between scientific inputs into policy making and democratic control is not a zero-sum game, the manner in which expertise is organised for policy making can be decisive for the scope and nature of public involvement.

In Denmark, the scientific community rarely appears in the public as a unitary actor, and it has not been a particularly vocal participant in the public debate on GMOs. Some academic researchers are committed to the promises of GM protagonists, but a significant proportion of public research funds have been channelled into risk research, which represents a more cautious approach in public. As such, there is no organisational locus claiming to talk publicly on behalf of 'Science' in these matters in Denmark. Consequently, there is less propensity to appropriate 'Science' as a unitary source of authority when compared to the UK or Germany. Therefore, the expert recommendations flowing into policy discussions have been less politicised and treated as relatively neutral representation of the best available knowledge. In relation to the coexistence issue, it has not been required that the public should adjudicate among competing knowledge claims, as it is often seen when experts disagree in public.

Although the political establishment in Britain is keenly aware that scientific expertise affiliated with Government received some serious dents in their public credibility during the BSE scandal, the British government is much more adamant that policies must be based on 'the best available science' and rigorous scientific risk assessment (DEFRA 2004). The official British position represents a stronger reliance on scientific knowledge as a means to settle controversies about risk regulation. However, even the attempts to compile broadly constituted and independent experts commissions have been met with accusations of a pro-GM bias. Critics claim that there is an alliance between dominant parts of the academic community and the biotech industry.[2] For that reason the scientific branch of 'GM Nation?' had to devote a significant effort to ensure its own credibility by publicly displaying even-handedness in appointment of experts for its review (Hansen 2010). In this situation public display of expert disagreement may break a technocratic monopoly of the 'facts', thus broadening the scope for meaningful interventions from the public. On the other hand, it does not necessarily enhance the agency of 'ordinary citizens' if they simply become spectators of a scientific controversy, which may not hold the answers or hold only partial answers to their concerns (Wynne 2001).

In Germany, the scientific community is comparatively better organised *vis-à-vis* the political system than in both Denmark and the UK. The German Science Foundation (DFG) is usually considered a unitary voice of 'Science' with a high standing in political debates. DGF has chosen a clear pro-GM stance (e.g. DFG 2004, also Union 2004) and the sceptics consequently reject mainstream science on the ground that it is in bed with industry interests. Instead, the sceptics rely on their own 'counter-expertise' drawn from the (less well funded) fringes of academia and 'alternative' research institutes, notably the Öko Institute (www.oeko.de). As a consequence, scientific knowledge has a much more contested status in the German debates than is the case in Denmark and to some extent also in the UK. Expert pluralism and disagreement is seen as an unavoidable state of affairs. This was particularly visible in the round table procedure (see above) where the ideal of impartial expertise searching for consensual interpretations of facts was abandoned from the outset and replaced with staged confrontations of different scientific viewpoints (Hansen 2010). These displays of expertise were more or less decoupled from the scientific advice actually flowing into policy making, a kind of separation not seen in the scientific review conducted in connection with the UK 'GM Nation?' process, which was controversial, yet served as a reference for subsequent policy formation (Beckett 2004).

Role of the public in implementing coexistence regulation

In Denmark, the political system was comparatively well aligned with public concerns when the coexistence question emerged on the policy agenda. The political climate therefore was less charged compared to the other cases, exhibiting less mobilisation and critical press coverage. Therefore, policy makers were able to rely on more conventional channels of public inputs – an expert hearing and a written consultation, which were both dominated by a strong technical focus on the efficiency of different regulatory means rather than debate

---

2   One example of this was a member of the GM scientific review carried out in connection to the GM Nation? Debate (see Horlick-Jones et al. 2007, Hansen 2010), Carlo Leifert, who resigned in protest against his perception of a strong pro-GM bias and strong industry influence on the review panel's work (Quoted in *The Guardian*, 24 July, 2003).

about the ends as previously seen in Denmark (Lassen and Jamison 2006). The avenues for engagement by a broader public were more limited than has been seen previously in Denmark, perhaps because the issue has (temporarily?) lost some of its potential to generate controversy.

In the UK in contrast, the general controversy on GM crops was still ongoing when the coexistence issue emerged on the policy agenda. On the one hand this has forced the political system to open up more arenas for articulation of concerns by the broader public along with stakeholders. On the other hand, the UK political system appears to have been much more immune to the substance of the concerns articulated. Rhetorically, the Government has remained adamant that only scientific assessments should be decisive for how the coexistence regime should be implemented (DEFRA 2004). In effect, public concerns have appeared in the eyes of the political system as the manifestation of unfounded opposition that must somehow be overcome through political tactics, not as a source of potential insights as the literature on participation oftentimes suggest – and which at certain moments seem to have been accepted by parts of the UK political system (HoL 2000).

The German case exhibits some more corporatist traits in the approach to public inputs into policy-making. Therefore the avenues of public engagement run almost exclusively via the established organisational landscape. Yet, the political leverage of organised stakeholders depends at least in part on the resonance they can generate for their viewpoints in the public sphere. In this struggle for public support the GM sceptics have been the most successful so far. They have also been less willing to compromise and maintained a fundamental opposition to GM agriculture, which is more radical than in the Danish case and more influential than in the British case, notwithstanding the fact that some GM crops have actually been planted in Germany. In this struggle among organised stakeholders for influence, the general public is offered limited scope to articulate concerns independently of existing organisational positions, which tends to disfavour more moderate positions when compared to the other cases (Hansen 2010). It thus seems that there is an institutionalised propensity to reproduce controversies over bio-objects in Germany, which renders fruitful public engagement in this area difficult (ibid.). The previous points of comparison are summarised in Table 6.1.

Conclusion

The three countries compared here have all carried out one or more national-level participatory exercises in order to engage the wider public with GMO issues prior to dealing with the coexistence issue (Hansen 2006b). Also in regard to the specific question of coexistence, which initially was thought to be only a technical detail, policy makers found it pertinent to solicit various kinds of inputs from stakeholders and to some extent the wider public in order to strengthen the legitimacy of the regulatory regime. However, a closer examination reveals that public engagement with the politics of bio-objects inevitably plays out in policy arenas with particular

Table 6.1   Public engagement and the governance of bio-objects in three countries

	Denmark	UK	Germany
Experiences with public engagement with GMOs	Longstanding, repeated engagement activities centred around the Danish Board of Technology	Significant change of policy rhetoric about public trust after BSE – e.g. the large-scale public debate 'GM Nation?'	Longstanding divisive evaluation of GM. Controversies addressed primarily in corporatist modes of governance
Discursive lines of contention regarding coexistence regulation	Between Government and anti-GM NGOs about the means of regulation	Between Government and general public about the ends of regulation	Internal controversies within parties in Government
Organising science for policy	Scientific committee considered as neutral source of knowledge	Attempts to constitute broad scientific advice in one organisational locus (but accused of pro-GM bias)	Scientific assessment divided between mainstream and alternative organisations competing for political attention
Role of the public in implementation of coexistence regualtion	Little wider public debate Written consultation and hearing at parliament	Coexistence regulation taken as occasion to manifest rejection of GM technology	Public concerns manifested through existing – polarised – organisational landscape and internally divided political parties

histories and established actor configurations. This significantly shapes the spaces that can be carved out for contributions from the public (see also Gottweis and Petersen 2009). While beyond the scope of explanation here, these differences illustrate that the calls for and analysis of processes of expanded public engagement need to pay close attention to the specific politico-cultural contexts in which such bio-objects are introduced in order to understand both the opportunities and limitations offered by public engagement with their governance. An important implication of this argument is that much more *systematic comparative research* is required in regard to public engagement with science and technology as a means to bring out the significance and effect of such contextual features.

As this chapter has hopefully demonstrated, the concept of 'bio-objects' constitutes a fruitful analytical tool for such comparisons. It can facilitate an understanding of how novel biological entities (and different kinds of knowledge of such entities) take on different meanings and generate different social dynamics depending on the contexts they are introduced into, for instance different political cultures, normativities or modes of commercialisation. As such, bio-objects also become co-constitutive of the 'publics' that are confronted by and engaging with them. Just as bio-objects can be said to have a fluid and transitory character, so do the publics of bio-objects. This makes the governance of bio-objects a particularly challenging task for which stable and robust routines still need to be developed.

References

AEBC 2003. *GM Crops? Coexistence and Liability*, Report from the Agriculture and Environment Biotechnology Commission. Available at: http://www.aebc. gov.uk/aebc/coexistence_liability.shtml [accessed 21 March 2010].

BDP 2007. Bundesverband Deutscher Pflanzenzüchter, *Bundeskabinett beschliesst Eckpunkte für neues Gentechnikgesetz* (press release. Available at: http://www. google.dk/url?sa=t&source=web&ct=res&cd=1&ved=0CAYQFjAA&url=htt p%3A%2F%2Fwww.gmo-safety.eu%2Fpdf%2Faktuell%2Fgentg_eckpunkte_ bdp.pdf&ei=MACqS5f0IsL5-Aa-zM1m&usg=AFQjCNG1vZdFgh6GuFuvdt MsJdfm5B_DjQ) [accessed 22 March 2010].

Beckett, M. 2004. 'Secretary of State Margaret Beckett's statement on GM policy', delivered to the House of Commons, 9 March 2004. Available at: http://www. rothamsted.bbsrc.ac.uk/pie/sadie/reprints/beckett_announcement_9_march.pdf [accessed 11 February 2009].

BIOSAM 2004. *GMO til Danmark*, BIOSAM informerer, no. 23. Available at http://www.biosam.dk/biosam/PDF/Biosam23.pdf) [accessed 21 March 2010] biosicherheit.de 2007. Available at: http://www.biosicherheit.de/de/aktuell/551. doku.html [accessed 22 March 2010].

BÖLW (Bund Ökologische Lebensmittelwirtschaft) 2009. *Gentechnik-Eckpunktepapier der Regierung: Eine Lizenz zum Verschmutzen*, press release. Available at: http://www.google.dk/url?sa=t&source=w eb&ct=res&cd=1&ved=0CAgQFjAA&url=http%3A%2F%2Fwww. boelw.de%2Fuploads%2Fmedia%2FPM_0607_Eckpunktepapier_ Gentechnikgesetz_070228_01.pdf&ei=8gCqS9qXPM2E-Qb1z8h- &usg=AFQjCNF5cUJU5isnpFxTnYnb2z7x_lZJIA [accessed 22 March 2010].

DEFRA 2004 (Department of Environment, Food and Rural Affairs). *The GM Dialogue: Government Response*, DEFRA, London.

DEFRA 2007 (Department of Environment, Food and Rural Affairs). *Summary of Responses to Defra Consultation Paper on Proposals for Managing the Coexistence of GM, Conventional and Organic Crops*, DEFRA, London.

DFG 2004 (Deutsche Forschungsgemeinschaft). *Stellungnahme der Deutschen Forschungsgemeinschaft zum Entwurf eines Gesetzes zur Neuordnung des Gentechnikrechts*. Available at: http://www.google.dk/url?sa=t&source=w eb&ct=res&cd=1&ved=0CAYQFjAA&url=http%3A%2F%2Fwww.dfg. de%2Fdownload%2Fpdf%2Fdfg_im_profil%2Freden_stellungnahmen%2F2 004%2Fgentechnikrecht_0604.pdf&ei=YgaqS6XNGMTc-QbWubBp&usg= AFQjCNF4wqw8Yq6lB5Qet9SwbV47ZCiHhg [accessed 22 March 2010].

Durant, J. 1999. Participatory technology assessment and the democratic model of the public understanding of science. *Science and Public Policy*, 26(5), 313-19.

European Commission 2002. *Life Sciences and Biotechnology – A Strategy for Europe* (Com (2002/C 55/03)), Brussels. European Commission 2007. *Taking European Knowledge Society Seriously* (EUR 22700), Brussels.

Fischer, F. 2000. *Citizens, Experts, and the Environment*. Durham, NC: Duke University Press.

Funtowicz, S.O. and Ravetz, J.R. 1993. Science for the Post-normal Age. *Futures*, 25(7), 739-55.

Gaskell, G., Allum, N., Wagner, W., Hviid Nielsen, T., Jelsøe, E., Kohring, M. and Bauer, M. 2001. In the public eye: Representations of biotechnology in Europe, in *Biotechnology 1996-2000: The Years of Controversy*, edited by G. Gaskell and M. Bauer. London: Science Museum, 53-79.

Gaskell, G., Allum, N. and Stares, S. 2003. *Europeans and Biotechnology in 2002 – Eurobarometer 58.0* (2nd Edition), Brussels: EC Directorate General for Research.

Gaskell, G. and Bauer, M. 2001. Biotechnology in the years of controversy: A social scientific perspective, in *Biotechnology 1996-2000: The Years of Controversy*, edited by G. Gaskell and M. Bauer. London: Science Museum, 3-11.

GeneWatch 2006. *GeneWatch UK Response to DEFRA Consultation on Proposal for Managing the Coexistence of GM, Conventional and Organic Crops*. Available at: http://setup.greennet.org.uk/uploads/f03c6d66a9b354535738483c1c3d49e4/ DEFRA_consultation1006.doc [accessed 22 March 2010].

Gottweis, H. 2008. Participation and the new governance of life. *BioSocieties*, 3, 265-86.

Gottweis, H. and Petersen, A. 2009. *Biobanks: Governance in Comparison*. London: Routledge.

Greenpeace DK 2004. *Regering erkender at GMO-samekistens er umuligt og spredning er uundgåeligt*. Available at: http://www.google.dk/url?sa =t&source=web&ct=res&cd=1&ved=0CAYQFjAA&url=http%3A%2F %2Fwww.greenpeace.org%2Fraw%2Fcontent%2Fdenmark%2Fpress% 2Frapporter-og-dokumenter%2Fhoringssvar.pdf&ei=XAOqS5qNCYjt-AbripiCAQ&usg=AFQjCNH-Sq0MfrSwFECwe9enrmN8TDE69w [accessed 22 March 2010]

Hansen, E.K. 2006a. "GMO i landbrug – vejen frem", speech delivered by the Danish Minister for Food, Agriculture and Fisheries at the EU conference on GMO coexistence, Vienna, 5 April 2006. Available at: http://www.fvm.dk/ Nyhedsvisning.aspx?ID=18486&PID=169609&NewsID=4569 [accessed 11 February 2009].

Hansen, J. 2006b. Operationalising the public in participatory technology assessment: a framework for comparison applied to three cases. *Science and Public Policy*, 33(8), 571-84.

Hansen, J. 2010. *Biotechnology and Public Engagement in Europe*. Basingstoke: Palgrave Macmillan.

Hagendijk, R. and Irwin, A. 2006. Public deliberation and governance: Engaging with science and technology in Contemporary Europe. *Minerva*, 44(2), 167-84.

Horlich-Jones, T., Walls, J., Rowe, G., Pidgeon, N., Poortinga, W., Murdock, G., O'Riordan, T. 2007. *The GM Debate: Risk, Politics and Public Engagement.* London: Routledge.

House of Lords. 2000. *Science and Society* (3rd Report of the Select Committee on Science and Technology). Available at www.publications.parliament.uk

Irwin, A. 2006. The politics of talk: Coming to terms with the 'new' scientific Governance. *Social Studies of Science*, 36(2), 299-320.

Jasanoff, S. 2005. *Designs on Nature*. Princeton, NJ and Oxford: Princeton University Press.

Jelsøe, E., Lassen, J., Mortensen, A.T. and Kamara, M.W. 2001. Denmark: The revival of national controversy over biotechnology, in *Biotechnology 1996-2000: The Years of Controversy*, edited by G. Gaskell and M. Bauer. London: Science Museum, 157-71.

Joss, S. and Bellucci, S. (eds) 2002. *Participatory Technology Assessment: European Perspectives.* London: Centre for the Study of Democracy, University of Westminster.

Lassen, J. and Jamison, A. 2006. Genetic technologies meet the public. *Science, Technology and Human Values*, 31(1), 8-28.

Levidow, L. and Boschert, K. 2008. Coexistence or contradiction? GM crops versus alternative agricultures in Europe. *Geoforum*, 39(1), 174-90.

Marris, C. and Joly, P-B. 1999. Between consensus and citizens: Public participation in technology assessment in France. *Science Studies*, 12(2), 3-32.

Mejlgaard, N. 2009. The trajectory of scientific citizenship in Denmark: Changing balances between public competence and public participation. *Science and Public Policy*, 36(6), 483-96.

Nowotny, H., Scott, P. and Gibbons, M. 2001. *Re-Thinking Science*. Cambridge: Polity Press.

Renn, O., Webler, T. and Wiedemann, P. (eds) 1995. *Fairness and Competence in Citizen Participation: Evaluation Models for Environmental Discourse.* Dordrecht, Boston and London: Kluwer Academic Publishers.

Rowe, G. and Frewer, L. 2000. Public participation methods: A framework for evaluation. *Science, Technology & Human Values*, 25(1), 3-29.

Sclove, R.E. 1995. *Democracy and Technology*. New York and London: The Guilford Press.

Süddeutsche 2009. *Deutschland verbietet Anbau von Genmais*. Available at: http://www.sueddeutsche.de/wissen/337/464931/text/ [accessed 22 March 2010].

Toft, J. 2007. Denmark's regulation of agri-biotechnology: Co-existence bypassing risk issues. *Science and Public Policy*, 32(4), 293-300.

Togersen, H. et al. 2002. Promise, problems and proxies: Twenty-five years of debate and regulation in Europe. *Biotechnology – The Making of a Global Controversy*, edited by M. Bauer and G. Gaskell. Cambridge: Cambridge University Press, 21-94.

Udredningsgruppen 2003. *Rapport fra udredningsgruppen vedrørende Sameksistens mellem genetisk modificerede, konventionelle og økologiske afgrøder*, Danish Ministry of Agriculture. Available at: pdir.fvm.dk/Admin/Public/DWSDownload.aspx?File=%2fFiles%2fFiler%2fVirksomheder%2fFro e%2fPlantegenetik%2fRap_GMsameksistens_082003.pdf [accessed 21 March 2010].

Union (Union der Deutschen Akademien der Wissenschaften) 2004. *Offener Brief und Memorandum zur Grünen Gentechnik in Deutschland*. Available at: http://www.google.dk/url?sa=t&source=web&ct=res&cd=1&ved=0CAgQFjAA&url=http%3A%2F%2Fwww.akademienunion.de%2F_files%2Fmemorandum_gentechnik%2Fmemorandum_gruene_gentechnik_offener_brief.pdf&ei=Qf-pS-CyAsGQ-Abp6fxo&usg=AFQjCNG3xfGWbQcpKQCmKs-OL8miSQjhFA [accessed 22 March 2010].

Verbraucherzentrale (Verbraucherzentrale Bundesverband) 2007. *Scharfe Kritik an den Eckpunkten zum Gentechnikrecht*. Available at: http://www.google.dk/url?sa=t&source=web&ct=res&cd=1&ved=0CAgQFjAA&url=http%3A%2F%2Fwww.vzbv.de%2F2Fstart%2F2Findex.php%3Fbereichs_id%3D%26mit_id%3D848%26page%3Dpresse%26task%3Dmit%26themen_id%3D&ei=vAGqS9P3I4rQ-QbH4uxq&usg=AFQjCNHIZLA6uNteI2Vpa 2zl8C0vQ1EtrQ [accessed 22 March 2010].

Vintersborg, K. and Pedersen, S. 2007. *Coexistence of Genetically Modified, Conventional and Organic Crops in the Nordic Countries*. Copenhagen: Nordic Council of Ministers.

Wynne, B. 2001. Creating Public Alienation: Expert Cultures of Risk and Ethics on GMOs. *Science as Culture*, 10(4), 445-81.

Zukunftsstiftung (Zukunftsstiftung Landwirtschaft) 2004. *Ein offener Brief und eine Einladung zum Streitgespräch*. Available at: http://kritischer-agrarbericht.de/fileadmin/Daten-KAB/KAB-Debatte-2004/winnacker_brief.pdf [accessed 22 March 2010].

Chapter 7

Governing Hereditary Disease in the Age of Autonomy: Mutations, Families and Care

Aaro Tupasela

During the last decade the role of molecular biology has made great strides in the identification and characterization of the molecular causes of a number of diseases. This process has led to what some commentators have described as the molecularization of medicine whereby disease, its identification and treatment is increasingly understood through the molecular model of life (de Chadarevian and Kamminga 1998). Molecules, in this sense, come to play an important role in the way we relate to disease, its treatment and management. In this sense molecules and the information that they carry can be said to become a type of bio-object to which various forms of (self) governance can be attributed, but more importantly in the ways it makes life matter. Life, in this sense, is tied to the materiality of molecules, but at the same time the prospect of living and maintaining life through the molecular model is brought back to the centre of care-taking (some might argue that care-taking has never left).

Novas and Rose (2000: 485) have argued that the creation of the genetically "at risk" patient through the molecularization of disease has given rise to new forms of active relations to one's self and one's future. They note that "when an illness or a pathology is thought of as genetic, it is no longer an individual matter. It has become familial, a matter both of family histories and potential family futures" (Novas and Rose 2000: 487). Rose (2007: 4) further argues that the evolution of marketization, autonomization and responsibilitization gives rise to a new contemporary politics of life itself.

In this chapter, I investigate the relationship between the molecularization of hereditary colorectal cancer and forms of care that are associated with people and families who have been identified as being at risk of developing the condition. Although we may have come to understand the aetiology of hereditary cancer through the molecular model, whereby mutations in DNA come to represent levels of personal risk, the treatment, management and most importantly, the care practices associated with hereditary cancer help to elucidate the ways in which the mutations (associated with hereditary cancer) as bio-objects are intimately associated and intertwined with practices and processes of care and treatment, as well as the technologies, such as databases, which are developed *in lieu* of their management. Molecularization, it is argued here, become intimately intertwined with a host of practices and technologies through which disease and patients are managed. In this sense I see bio-objects as having a transformative power in practices of care.

Recent research on the forms of governance and risk management imposed by new high-tech medicine has suggested that what has emerged as a result of the molecularization of disease is a form of existential responsibility whereby "high-tech biomedicine tends to individualize risks and impose a form of individuality characterized by the demand for reflexivity through personal risk assessment" (Helén 2004: 28). Although recent medical discourse surrounding patient choice has come to emphasize the autonomy of the patient, it is through practices of treatment, I argue, that this notion can be problematized. Individualization, I would like to argue, is intertwined with a host of practices and technologies which make the management of disease a much broader matter. In this sense the dichotomy between autonomy and paternalism which is set forth in medical sociology and bioethics literature becomes blurred and challenges our understanding of how personal risk is considered within this context.

In light of this, I want to draw connections between the genetic mutations associated with hereditary cancer, as a bio-object, how this is understood in the research setting and the way this is related to how hereditary cancer is treated and governed in the health care sector. This linkage allows us to examine the way in which knowledge of hereditary cancer allows for new forms of intervention and prevention, but also the ways in which it challenges our understanding of privacy and autonomy. With hereditary cancer the impact of the knowledge of the molecular causes of the disease has bearing on not just the individual patient and their families, but more importantly on a whole system of care that is associated with the research, diagnosis and treatment of the condition. In this sense, hereditary disease as a bio-object extends our scope of investigation beyond the individual and their families and relates it to whole systems of research, care and treatment that are made and become available to them.

At the heart of this relationship between research and care lies an imperative, an *imperative for prevention*. I use the word imperative for prevention for a number of reasons. In placing the governance of hereditary disease within a broader political context, I draw attention to the relation between the Nordic welfare state, health care and biomedical research. I see this as an important relationship in that traditionally the Nordic welfare state has had the responsibility to provide services and guide health care activities. Despite health care increasingly being privatized, the state retains the main responsibilities in providing health care to its citizens. At the same time, the state maintains the main responsibility in funding biomedical research and has stated that it is its intention to fully utilize the benefits that are gained as a result of that research. In this sense, the results of research are channelled into health care, regardless of whether they are in the public or private sphere (Tupasela 2008).

I would also like to draw on recent work done on choice; in particular choices that are made and become available during the course of care and treatment (Mol 2008). I do this for several reasons. *First*, the notion of choice has been argued to be seen as a mainstay within the medical field, whereby patients are offered choices from which they can choose. *Second,* choice relates to the notion of autonomy, whereby individuals are seen as independent actors. Yet with the diagnosis and treatment of hereditary cancer, the management of hereditary disease is intertwined within a much broader network of actors and processes which relate to the management of hereditary cancer within whole families. *Third,* the notion of choice draws our attention to the actions of individuals within a much broader setting, namely the health care sector, as opposed to independent individuals operating outside of any context. *Fourth,* the molecularization of disease has in some ways been understood in relation to the disease being a material bio-object; in this case the mutations that are associated with the hereditary condition. I would like to, however, re-direct attention to the practices and technologies that surround such bio-objects and the ways in which care and treatment practices emerge in the processes of managing and dealing with such facts.

Recently, the idea of patient choice has been re-examined and questioned in terms of its relationship to forms of care and treatment that are made available to patients. Some commentators have noted that the idea of non-directiveness is difficult to achieve in situations where there is an imperative for prevention (Nordahl Svendsen 2006; Koch and Nordahl Svendsen 2005). Mol, for instance, has noted that "introducing patient choice into health care does not (finally) make space for us, its patients. Instead, it alters daily practices in ways that do not necessarily fit well with the intricacies of our disease" (Mol 2008: 2). The relationship between choice and care is, therefore, not a straightforward one, but one where the boundary between our choices and the available forms of care and treatment intersect in ways that challenge our traditional assumptions surrounding patient autonomy and choice. In this sense the processes of care associated with the bio-objects of molecular mutations that cause hereditary cancer involves also shifts in the way diseases are managed and governed. In this sense bio-objects are not just physical entities (genetic mutations), but rather intertwine with socio-temporal assemblages of patients, families and the care givers associated with their care (see Douglas in this volume), as well as national systems of health care.

Hereditary cancer and care

The 'duty to warn' a patient's family members of hereditary disease risks has raised a number of legal and ethical discussions concerning the dilemma of doctor's "obligations to respect the privacy of genetic information vs. the potential legal liabilities resulting from the physician's failure to notify at-risk relatives" (Offit et al. 2004: 1469). With the development of molecular biology, information and hereditary diseases have begun to pose an increasing problem in terms of decisions to inform or not to inform patients and their relatives who might be affected and if the decision is made to inform, then in what way should it be done. Such tensions are informed by two traditions. The first, where autonomy and non-directiveness are emphasized and individual choice ought to be supported by non-directive counselling. Medical professionals are expected to provide value-neutral information of risks and subjects of counselling are expected to make independent and rational choices (Petersen 1999). Eriksson (2004: 46) has suggested that a strategy where results are disclosed should be discussed already during the informed consent phase, but in cases where this has not been done, there should be an attempt to strengthen 'the subjects' autonomy and encourage them to take responsibility. This, it is argued, would avoid medical paternalism, and respect the autonomy of research subjects. Similarly, Koch and Nordahl Svendsen (2005) have proposed that the imperative of prevention in genetic counselling is changing the notions of directiveness. Although the practices surrounding genetic subjecthood and the dissemination of information concerning hereditary disease are changing, the relationship between research practices and patient care through prevention remain a challenge.

New genetic knowledge has brought with it a responsibility to act, both for patients and physicians. For patients new biomedical knowledge on personal risks opens up the possibility to act and intervene on one's own body. In the case of hereditary disease this also extends to the family as well. For physicians and other professionals, responsibility also entails imparting knowledge in a way that is not coercive or directive. Warning family members of a life-threatening inherited condition highlights the challenge to try to respect privacy on the one hand, but also to protect lives on the other. It also highlights the ambiguity that emerges for professionals in the role that they are supposed to take: non-directive on the one hand, yet responsible on the other. Without pre-existing legal frameworks physicians are unsure of the ways in which and through what types of processes they are expected and should intervene and treat patients.

In the rest of this chapter, I will look at research carried out on Hereditary Non-Polyposis Colorectal Cancer (HNPCC or Lynch Syndrome). Drawing on interviews with medical researchers in Finland who have studied the condition in Finnish families for decades and who have developed a screening and check-up process to detect the onset of the condition early on in order to allow for early intervention I examine the ways in which hereditary mutations as bio-objects have transformative power altering the way we relate and act upon those who are cared for. Early intervention through surgery has been proven to be the most effective way of preventing the spread of the cancer and continued interaction between the patients and the hospital is an essential part of the treatment procedure (Järvinen and Mecklin 1994). At the same time, however, since the cancer is hereditary the treatment process of individual patients also entails further contacts with other family members who may be at risk.

Research in Finland began in the early 1980s to identify families who were carriers of the condition. The research ultimately led to the identification of 107 families or over 300 people who carry the mutation and are at risk. During the course of the research, the researchers developed a database of all the family members and relatives who were and might be at risk so that they could be called up regularly for screening. The screening allowed for early intervention through surgery. The problem which arose, however, was how should those family members who were at risk, but did not know about it, be contacted and informed of their risk. In order to protect patient privacy of those family members who had been identified, the doctors could not tell the other family members through whom their suspicion had emerged. At the same time, it was difficult and challenging to directly inform the relatives of patients that they must come in for check up's since this would be too directive.

I frame the issue of governance of hereditary disease within a broader bio-political pre-occupation related to care giving and an imperative for prevention. By doing this I seek to extend our understanding of the relationship between bio-objects, as genetic mutations which cause hereditary conditions, to relations of care giving and treatment between researchers and patients. What is at stake here is the tension between two perspectives on the governance of life. On the one hand, the governance and management of life is allocated to the individual/ the patient/the self, whereby individual choice serves as the mechanism through which life and vitality is controlled and maintained. On the other hand there is a state imperative to save lives, where the administration and care associated with vitality and life is managed and dictated by others, namely doctors and various state institutions. This type of medical paternalism has come to have a negative connotation where experts exert their will and wish on patients through unequal power relations with patients. Individual choice and autonomy are seen to remedy this problem by shifting responsibility to the individual and recent developments in medicine suggest that there is a need to strengthen the patient's possibility to make such autonomous decisions. The case of HNPCC research points, however, to inadequacies that the dichotomy between autonomy and paternalism has in relation to the way new bio-objects are challenging the way care is delivered.

The logic of research and care

The tension between the autonomy of research subjects/patients and medical paternalism has to be understood within the context in which the molecular knowledge surrounding the diagnosis of the hereditary condition came about. The identification of the mutations and their translation into treatment are closely aligned with practices of both research and care. Work on HNPCC can be divided into two general areas: one comprising the clinical work, where two doctors at different university hospitals in Finland coordinated their work directly with patients and families, providing regular check-ups, as well as treatment; and the genetic research component where researchers worked to discover the genetic causes of the mutation in the families, using records and samples of patients who were not part of the clinical research component.

Within the research context the research team were able to compile a database of all those who took part in the research, as well as develop family histories to identify those not yet aware that they might be at risk, but who should be contacted

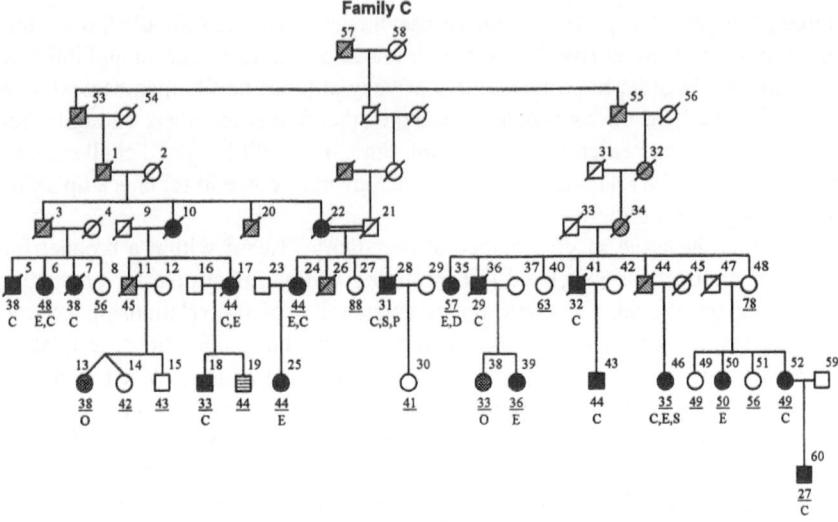

Figure 7.1   Pedigree of family predisposed to hereditary cancer

Source: Peltomäki et al., 1993, reproduced with permission.

since they were at risk of being not only carriers, but also of developing the condition. These family pedigrees were an important part in developing a 'map' of all people in Finland who were at risk of carrying the genetic mutations that led to the condition (see Figure 7.1). The image represents seven generations of a family that is predisposed to hereditary cancer.

Within the context of the research setting the logic that operated was two-fold. On the one hand, the researchers were interested in identifying the genetic causes of the hereditary condition. On the other hand, they were also committed to providing preventive treatment to those that were participating in the research, as well as finding a way of helping those family members who were not yet aware that they were at risk. Contacting those family members who were not aware of their risk raised, however, the question of under what conditions could contact be made without undermining the privacy and autonomy of the patients who were being treated. At the same time, however, researchers were also motivated by the imperative for prevention whereby they felt obligated to ensure that as many people were contacted and warned as possible. The tension between the respect for privacy and the imperative for prevention was pivotal in the decision-making process of the researchers, but the practices associated with the compilation of the database also played an important part in the process in that they had to remain non-directive.

The database of family members that the researchers used was compiled within the research context whereby researchers are allowed by law to compile data on people based on disability. The logic that operates within the research setting is based on allowances in relation to the use and categorization of personal data which are less strict than those in the health care setting. The maintenance and use of such a database outside the research setting would not be possible, therefore, since database legislation prohibits the compilation of personal data based on disability or disease. As a result, while the research was being conducted the treatment and care that the research patients received with the help of the database was allowed by law only in the research setting. When the researchers finished their work and wanted to pass the care and treatment process over to the health care sector, problems began to emerge since the logic of care in the health care setting is based on a different interpretation of the way in which personal information can be managed, stored and used in the health care setting.

It is at the intersection of these two logics that the tension between personal autonomy and privacy is confronted by the preventive imperative within biomedical research. In the next section the relationship between these two logics and the resulting tension will be examined in order to critically analyse the notion of patient autonomy and its relationship to the molecularization of hereditary cancer.

Choice and care in hereditary cancer

During the course of conducting research on HNPCC the research team and its associated clinical arm became worried about the family members of those who had been identified with carrying the gene mutations. It was obvious that the ability to compile databases of family pedigrees that mapped out the gamut of those family members who might be at risk was a new problem that had arisen as a result of identifying not only the genetic mutations, but also being able to link this to family trees as well. At the same time, however, the researchers were aware that they were treading a thin line in relation to patient non-directiveness, privacy and autonomy. As a result, the researchers devised a way in which they would be able to get in touch with family members and open up a 'space' in which it would be possible for them to receive counseling, as well as gain access to predictive testing and ultimately treatment if need be. This required, however, an approach that was non-directive, but also provided enough information to the family members for them to realize that the issue at hand was possibly life threatening and important that they seriously consider the matter. Physicians, therefore, decided to go the route whereby they sought direct contact with family members who had not participated in the research and were not aware of the risk that they were in (Aktan-Collan et al. 2007). This approach represents a particular type of engagement with family members and relatives which sought to re-configure existing notions of privacy and autonomy in that the preventive imperatives of research extend beyond the original research patient.

The research team developed a two-tier approach whereby they would first contact people by mail and then by phone. The idea was to make people aware in some way that they might be at risk and that counselling, testing and treatment was available if they would want to accept it. The overall approach utilized by

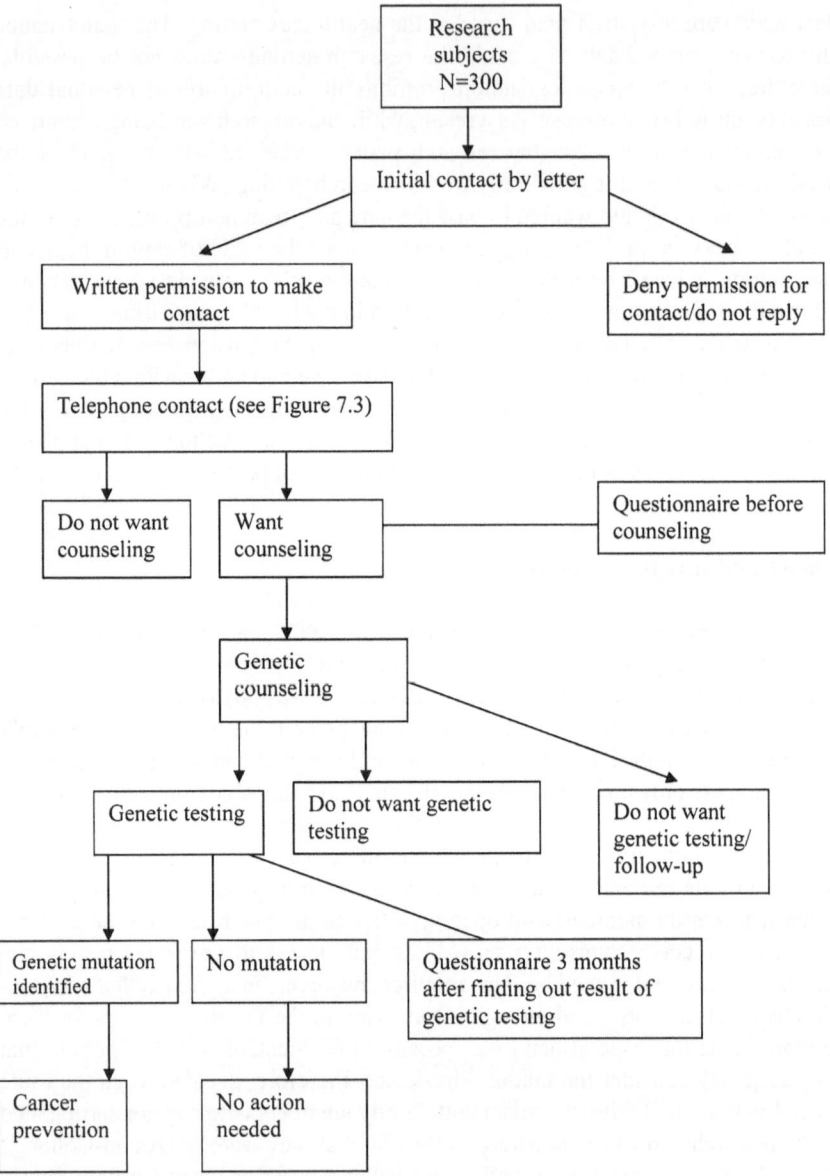

Figure 7.2   Contact protocol for possible HNPCC carriers

Note: The protocols pictured here have been described in and are based upon Aktan-Collan, K. et al. 2007.

the research team allowed for numerous opportunities for the people who were contacted to 'opt-out.' The main challenge of this approach was, however, that simply contacting the people and making them aware of the fact that they might be at risk was likely to cause distress and concern. As a result, the research team was also interested in studying the effect of the contact process on the people they contacted. In Figure 7.1, the overall schema for the contact process is detailed. This process was undertaken with 300 people who were considered to be at risk of being carriers of the condition.

Although some commentators have noted how high-tech medicine has retreated to a position of *technical* responsibility, leaving patients with the choice and ethical responsibility of what to do (cf. Helén 2004), I would like to provide a somewhat different perspective on this approach based on this contact schema; one that is rooted in an imperative to care and the responsibility of the researcher to make contact with patients.

Although it is clear from the contact schema that it is the patient who carries the ultimate responsibility to make the choice of whether or not to accept counselling, genetic tests and finally treatment, I argue that the fact that the research team has developed such a scheme for contacting family members suggests that notions of responsibility, risk and ethics can also be located within the medical community and researchers, as well in relation to how one should act in light of new knowledge based on the genetic mutation. The notion of bio-objects, however, takes on a much broader significance in this context. It is not just the genetic mutation, but also necessarily includes the genetic test for the mutation, the formulation of the pedigrees, the contact schemas used in providing care, as well as the database that has been compiled. In this sense, the notion of an individual carrier as a patient becomes intertwined with a whole host of processes of research, care and technologies which are necessary for the management of the disease.

The notion that bio-objects create forms of subjectivity among patients can be contrasted and compared to the forms of subjectivity that are created among the researchers who create this knowledge and the possibility for intervention as well. Questions of genetic privacy (Laurie 2002), information and hereditary diseases have begun to pose an increasing problem in terms of decisions to inform or not to inform patients and their relatives who might be affected. It can be argued that this position also generates a form of anxiety among medical professionals that is generated by new forms of responsibility.

This anxiety and the question of what actions to take is further illustrated by the phone interview framework through which family members who wanted contact to alert them of the possibility that they might be at risk (see Figure 7.2). The approach is very general at first, enquiring people whether they are in any way aware that someone in the family has had cancer at an early age. If people answer yes they are aware then they go on to inform them that it appears that they are in one of those families that is at an increased risk of being a carrier of one of the mutations. At the same time they are offered counselling if they would like to receive it since they might be at risk. It is through the counselling service that they are then directed to the counselling protocol which means that they are offered the possibility to have a genetic test to be done. If the test is positive then they are offered the possibility to participate in regular colonoscopy check-ups through which early detection of the lesions is made possible. The early detection leads to early intervention and thus a significantly higher possibility of survival.

For those who are not aware that there is cancer in the family the research staff takes a different approach in relation to talking about the possibility that the family member may be at risk. The notion of non-directiveness and autonomy is re-configured by asking the patient what they would like someone to do if they knew that they might be at risk. It is at this juncture that the relationship between the patient and care-giving becomes expressed in a way where the patient is able to maintain some semblance of autonomy and choice while at the same time allowing for the possibility of intervention through prevention. This position is a tenuous one since contacting family members via telephone and asking suggestive questions opens up the possibility within the family members' mind that something may be wrong. The notion that responsibility is transferred over to the patient is only partially true in that it is the medical research team that also maintains responsibility in relation to contacting the patients in the first place. This motivation to contact is driven by the imperative for prevention and responsibility for the lives of other people that has been made possible by the discovery of a bio-object that manifests itself as a hereditary condition, and thus touching the lives of other people besides simply that of the patient. It is at this juncture, I argue that the preventive capacity and the possibility to offer care outweigh notions of privacy and autonomy in such a way that researchers and clinicians see it to be their duty and indeed ethical responsibility to contact those who are at risk.

My point in raising this issue does not, however, relate directly to issues and concerns over privacy, autonomy and non-directiveness, but rather to the issue of care that is made possible through contact. The counselling that is offered to both those who have heard of the condition and of people in the family who might have suffered from it, as well as those who have not, is a central element that operates between the patients and their family members and the clinical researchers.

The theoretization related to the relationship between researchers and patients has focused too much on the power relation between the two operative sides as oppositional, where the actions of researchers and clinicians are seen through the lens of paternalism. Although the idea of care and taking care of someone has connotations related to paternalism, I would argue that in asking patients what they would do, the power relation is re-configured in relation to actions that ought to be taken in a given situation. Granted, this situation is 'directive' since it insinuates that there is clearly a possibility that a chance of cancer may exist, the idea that patients themselves come to define the scope of responsibility that they would see as appropriate in such a situation also reverses the idea that responsibility and authority in care taking is unidirectional.

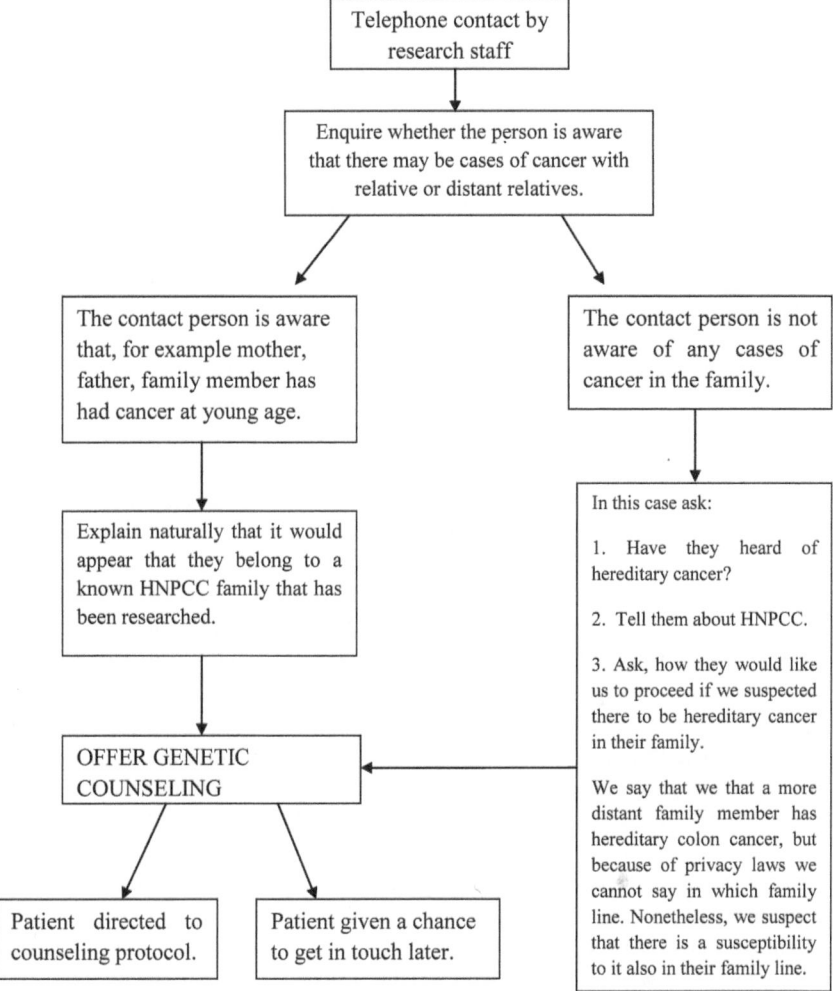

Figure 7.3   Telephone contact protocol for HNPCC family members and relatives

Note: The protocols pictured here have been described in and are based upon Aktan-Collan, K. et al. 2007.

Although recent discussions surrounding the emergence of the "somatic individual" (Novas and Rose 2000) have shifted our attention towards the ways in which *individuals* are increasingly taking responsibility for their lives and health through the mediation of genetic risk, the responsibility and initiative to organize counselling, treatment and operative care is still organized and offered by medical professionals. To term this organization of medical activities around the singular bio-object of hereditary genetic mutations as medical paternalism or a threat to autonomy is misleading since it does not take into account the significance of responsibility by both parties. At the same time, although it may seem that it is the patient who is taking responsibility for their own condition, I would argue that the responsibility is shared within the structure of the telephone interview and the subsequent treatment if the patient so chooses.

Mol (2008) has argued that the framing of patients as customers within recent years is problematic in relation to the notion and practices that are related to care. I would go even further and argue that the construction of the patient as customer is even more problematic in relation to the notion of responsibility towards one's own health within the context of the Nordic welfare system of health care in that within these systems, despite the fact that authorities are responsible in organizing and providing equal access to treatment, they are nonetheless seen to be obtrusive if they offer such services to people they may suspect are in danger of a life threatening condition. If the emergence of the responsible patient is premised on the possibility of making choices then certainly there should be choices that one is able to make. In the case of hereditary cancers, however, the medical establishment is not legally allowed to make a clear choice available to the 'customer', but rather insinuate that there may be a choice which the patient might want to consider. Choice is based, after all, on a notion that there is information available on which the consumer is able to make informed decisions. The logic which drives health care systems towards mechanisms that support choice among patients is hindered to a great extent by privacy laws.

Ironically, however, privacy laws operate only within the confines of health care, where choice is of utmost importance and not within the research system itself where there are exemptions as to the ways in which personal information can be managed and used. This curious dichotomy, in relation to the ways in which we manage information on people and their bodies, brings forth a set of problems related to the delivery of care in relation to hereditary cancers. Arguably one could say that this problem will be a thing of the past in a few decades when nearly all family members will have been contacted in one way or another. Yet the question of how the delivery of care to people who are at risk and need it the most has become such a contentious issue, ravelled up in ideological and philosophical questions related to privacy and autonomy, is an interesting one nonetheless. Perhaps, it is related to a discussion that has risen out of the molecularization of medicine, whereby genetic information is seen as 'special' or containing information that is more sensitive than other medical information.

In the case of HNPCC, doctors decided that the interests of individuals who were not aware of their conditions should outweigh the strict privacy issues of individuals who had participated in the research. The doctors, therefore, chose to contact family members and recommend counselling, as well as a DNA test that could help them determine if they were at risk (Tupasela 2006; Aktan-Collan et al. 2007). The 're-allocation' of responsibility to the patient by asking them what would they do is one way of getting around this issue, but in my opinion side-steppes an important problem within the health care system in relation to responsibility. It seems problematic to produce certain types of knowledge on the human body, on the one hand, but at the same time limit the possibilities of its use and application on the other. This tension in the role of the state as the main funder of research, as well as the implementer of health care policy becomes obvious. At the same time there appears to be a need to better understand the way forms of responsibility and ethical action figure among medical practitioners in relation to the discovery of new bio-objects. The re-interpretation of the limits of privacy and the extension of responsibility by physicians is important in relation to new bio-objects, such as hereditary genetic mutations, in that it asserts a type of moral responsibility and new form of engagement on behalf of researchers to act and intervene on the basis of the scientific knowledge that they produce.

Conclusion

Mol (2008: 89) has noted that "the logic of care itself is first and foremost practical. It is concerned with actively improving life". The notion that new high-tech medicine and the molecularization of disease has brought about new forms of responsibility is no doubt well founded and documented within the scientific literature. In this chapter, however, I have sought to introduce a more nuanced understanding and picture of the way responsibility and the ethical dimension surrounding bio-objects come to be configured as assemblages of care and treatment that are made available to patients and their family members. Within this framework forms of responsibility are shared, although in different ways, by medical practitioners and patients alike.

The forms of care and treatment that are made available to patients through the schemas of contact that I have outlined are by no means trivial, but rather central elements where doctors and patients alike share an anxiety over the possibility of serious disability or death and the ways in which it can be averted through practical means of intervention, prevention and treatment. It is in this sense that the notion of bio-objects, as it relates to hereditary mutations, comes to re-configure our understanding of how responsibility is shared among actors who are intertwined and brought together in one way or another by genetic mutations. The transformative power of science extends to a broader field of activity that is also outside the laboratory or clinical research setting. At the same time the tension between people as research subjects and patients is brought to the fore in light of processes where mutations as material bio-objects open up new possibilities of intervention.

References

Aktan-Collan, K. et al. 2007. Direct contact in inviting high-risk members of hereditary colon cancer families to genetic counselling and DNA testing. *Journal of Medical Genetics*, 44, 732-8.

de Chadarevian, S. and Kamminga, H. (eds) 1998. *Molecularizing Biology and Medicine: New Practices and Alliances, 1910s-1970s.* Studies in the History of Science, Technology, and Medicine, 6. Amsterdam: Harwood Academic Publishers.

Eriksson, S. 2004. Should results from genetic research be returned to research subjects and their biological relatives? *TRAMES,* 8(1/2), 46-62.

Helén, I. 2004. Technics over life: Risk, ethics and the existential condition in high-tech antenatal care. *Economy and Society,* 33(1), 28-51.

Järvinen H.J. and Mecklin J-P. 1994. Paksusuolensyövän seulonta. *Duodecim,* 110, 1919-24.

Koch, L. and Nordahl Svendsen, M. 2005. Providing solutions – Defining problems: The imperative of disease prevention in genetic counselling. *Social Science and Medicine,* 60, 823-32.

Laurie, G. 2002. *Genetic Privacy: A Challenge to Medico-Legal Norms.* Cambridge: Cambridge University Press.

Lynch H.T., Cristofaro G., Rozen P., Vasen H., Lynch P., Mecklin J-P. and St. John J. 2003. History of the international collaborative group on hereditary non-polyposis colorectal cancer. *Familial Cancer,* 28 (Suppl. 1), 3-5.

Mol, A. 2008. *The Logic of Care – Health and the Problem of Patient Choice.* Oxon: Routledge.

Nordahl Svendsen, M. and Koch, L. 2006. Genetics and prevention: A policy in the making. *New Genetics and Society,* 25(1), 51-68.

Novas, C. and Rose, N. 2000. Genetic risk and the birth of the somatic individual. *Economy and Society,* 29(4), 485-513.

Offit K., Groeger E., Turner S., Wadsworth E. and Weiser M. 2004. The 'Duty to Warn' a patient's family members about hereditary disease risk. *Journal of American Medical Association,* 292, 1469-73.

Peltomäki, P. et al. 1993. Genetic Mapping of a Locus Predisposing to Human Colorectal Cancer. *Science,* 260(5109; May 7), 751-2.

Petersen, A. 1999. Counselling the genetically 'at risk': The poetics and politics of 'non-directiveness'. *Health, Risk & Society,* 1(3), 253-65.

Rose, N. 2007. *The Politics of Life Itself: Biomedicine, Power, and Subjectivity in the Twenty-First Century.* Princeton, NJ: Princeton University Press.

Tupasela, A. 2006. When legal worlds collide. From research to treatment in hereditary cancer prevention. *European Journal of Cancer Care,* 15(3), 257-66.

Tupasela, A. 2007. Re-examining medical modernization – Framing the public in Finnish biomedical research policy. *Public Understanding of Science,* 16(1), 63-78.

Tupasela, A. 2008. *Consent Practices and Biomedical Knowledge Production in Tissue Economies.* Academic Dissertation. Department of Sociology Research Reports no. 256, University of Helsinki.

Chapter 8

At the Margins of Life: Making Fetal Life Matter in Trajectories of First Trimester Prenatal Risk Assessment (FTPRA)

Nete Schwennesen

This chapter traces the multiple ways in which fetal life comes into being in trajectories of first trimester prenatal risk assessment (FTPRA) for Down syndrome at an ultrasound clinic in Denmark and offers a reading of such practices as processes of bio-objectification.[1] In discussions about ethically fused practices on the border of new reproductive technologies (such as prenatal testing, abortion and stem cell research) a major concern has been where to draw a morally and ethically acceptable line between life and non-life on the basis of philosophical arguments, in order to legitimate or dissolve such practices (Holm 2005, 2000). Feminists have emphasised the pertinent political character of this line and the rights of women and fetuses are often set in opposition. In her study on fetal surgery the American anthropologist and science study scholar Monica Casper demonstrates convincingly that women's interests and fetuses' interests may contradict one-another in contemporary practice of reproductive technologies. In her study she critically examines how notions of the "human" are constructed in the socio-material practices of fetal surgery. Analyzing such practices, she argues that:

> Within the domain of experimental fetal surgery, the fetus is constructed as a potential person with human qualities. In weekly fetal-treatment meetings, for example, fetuses are routinely referred to as "the kid", "the baby" and "he" – all quite human (and gendered) attributions. (Casper 1994: 843)

On the basis of her case study Casper discusses the issue of nonhuman agency in relation to the debate on this issue among science studies scholars. Casper criticises the very notion of nonhuman agency for being premised on a dichotomous ontological positioning of the nonhuman as opposed to the human. By working

---

1 This chapter builds on the authors previous work which was made in fulfilment of a PhD project exploring the social implications of first trimester prenatal risk assessment in Denmark (Schwennesen et al. 2009, Schwennesen and Koch 2009, Schwennesen et al. 2010, Schwennesen 2011, Schwennesen and Koch 2012).

with pre-established categories of human and nonhuman agency a crucial factor from the analysis is excluded from analysis, since the attribution of 'human' and 'nonhuman' to entities is the consequence of particular political practices. On the basis of her analysis of fetal surgery, she warns that "constructions of active fetal agency may render pregnant women invisible as human actors and reduce them to technomaternal environments for fetal patients" (Casper 1994: 844). From this critical point, she argues against the attribution of fetal agency, which she justifies as follows, "my refusal to grant agency to fetuses, while simultaneously recognising it in pregnant women ... is about taking sides. My politics ... are about figuring out to whom and what in the world I am accountable" (Casper 1994: 853).

I share with Casper the argument that *a priori* distinctions between the human and the nonhuman might foreclose our ability to critically analyse the processes through which boundaries are drawn between the human and the nonhuman and how such boundary drawing always involves constitutive exclusions and hence may exclude possibilities of action. However, I see a problem with her simultaneous constellation of a universal boundary around fetal life, which she discards as having any form of agency. It seems that she ends up enrolling her work into the same dichotomy that she wanted to discard from the very outset. In the following I challenge Casper's feminist strategic position on fetal agency by way of offering a reading of fetal life as a contemporary form of bio-object. With inspiration from the feminist science study scholar Karen Barad's philosophical account of agential realism, I argue for a position that makes possible an account of fetal agency which simultaneously allows critical interrogations of how such boundaries are made, and what their effects are. In doing so I will present an analysis of how fetal life is made to matter, that is, how fetal life is bio-objectified in different ways throughout trajectories of FTPRA at an ultrasound clinic in Denmark.

First trimester prenatal risk assessment (FTPRA)

FTPRA is a combined risk assessment and screening technology carried out between weeks 11 and 13 of pregnancy. It involves an ultrasound scan and measures whether the fetus is at risk of having Down syndrome or other chromosomal diseases. If the risk is categorised as "high" according to a cut-off (1:300) the pregnant women and partners undergoing FTPRA will have to decide about whether or not to undergo technologies of invasive testing (CVS or amniocentesis) which are diagnostic but involve the risk of inducing a miscarriage. FTPRA is considered the most effective non-invasive screening technology for Down syndrome and other chromosomal disorders on the market (Nicolaides 2004), with an estimated detection rate of 85-95 per cent.

The outset of the study is the situation in Denmark, where FTPRA has been offered to every pregnant woman – regardless of age and risk – since 2004, on the basis of informed choice. Denmark is the first country in the world to offer

FTPRA free of charge to every citizen, regardless of age and risk situation. It is estimated that the current overall uptake is at least 90 per cent (Tørring et al. 2008). This study was conducted in the Copenhagen area where the uptake is estimated to be around 95 per cent (Tabor 2006). As such, FTPRA has become an almost obligatory passage point (Latour 1987) for pregnant women and partners in Denmark on their trajectories towards giving birth.

Investigating life as a process of mattering

At the heart of the question of fetal life, lies the question of materiality. The question is whether fetal life is to be considered an autonomous entity of human life and agency, or just a passive part of a pregnant woman's body. Ethical and feminist discussions on this issue often end up in dualistic accounts of the human and the nonhuman. In contrast the concept of bio-object, directs attention to the process through which different life forms are created and are given life. That is life is not considered a pre-given fact, but as something which is contingent and may emerge in different forms. The central focus therefore is the ways in which "life is made an object in different settings" and the myriad ways in which life "become an object of knowing, representing and intervening" (Introduction in this book). From the outset we are allowed to study bio-objects, in this case fetal life, as "having considerable fluidity and mobility across different socio-technical domains or arenas" (Introduction in this book). This approach reflects an unstable ontology, where ontology is conceived as an ongoing process rather than a stable form of being. Hence, the process through which certain ontologies becomes locally stabilised may be described as the outcome of a socio-material process, where entities are made stable in certain moments in time (Thompson 2005, Cussins 1998).

At the same time it is recognised that materiality has agency or performing capabilities. To clarify this matter, I suggest drawing on the feminist science study scholar Karen Barad's theory of agential realism (Barad 2007). Barad develops her theory of agential realism in conversation with, and as a contribution to, Judith Butler's rephrasing of the concept of performativity to involve the discursive processes through which bodies are materialised as sexed (Butler 1993). Barad criticises Butler for conceptualising the notion of performativity exclusively in discursive terms. Hereby, Barad argues, Butler comes to reinstall materiality (sex) as something which is outside discourse, and fails therefore to consider materiality as playing an active role in the constitution of gendered bodies. In other words, being excluded from analysis is how matter comes to matter in such processes.

Barad offers a materialist elaboration of performativity – a philosophical account that she calls agential realism – that calls into question the taken-for-grantedness of differential categories, such as human and nonhuman while simultaneously allowing "… matter its due as an active participant in the worlds ongoing becoming" (Barad 2008: 122). In doing so, she develops the concept of phenomena, with inspiration from the physicist Niels Bohr. Her starting point is his demonstration that classical correspondence theories of scientific knowledge fall short when trying to explain the classical wave-particle paradox; that light manifests particle-like properties under one set of experimental circumstances and wave-like properties under a different set of experimental circumstances (Barad 2007: 198, Bohr 1958). To a classical realist who sees the relationship between scientific knowledge and the object under observation as a corresponding or mirroring relationship, this situation is paradoxical: "the true ontological nature of light is in question: either light is a wave or it is a particle, it cannot be both" (Barad 2007: 198). Bohr resolves this problem, in Barad's interpretation, by installing "phenomena" understood as the wholeness of the entity of observation and the apparatus of observation as the referent of knowledge instead of the duality of an independent observer and an observation-independent object. In Barad's terminology, apparatuses of observation are not merely observing instruments but material-discursive practices through which differential boundaries are drawn (Barad 2007: 140). Barad emphasises that apparatuses of observation are themselves phenomena in the making, only temporally engaged in and articulating such boundaries (Barad 2008: 140). Barad's point is that the specific configuration of apparatuses of observation we use to make entities knowable should not be framed as passive or innocent (Barad 2007: 33). Rather, they should be conceived as productive and part of the phenomena, which emerge through processes of knowledge production.[2]

If we apply this framework on the practices of FTPRA, we should understand the fetus as a phenomenon which "includes" the whole configuration of apparatuses of observation, which is brought about in the production of knowledge about what the fetus "is" (such as visual images, risk figures, the pregnant woman, her partner, the sonographer, discursive interpretations etc.). As such, who or what the fetus becomes (human/nonhuman, low risk/high risk) in FTPRA processes of bio-objectification is not given in the fetus itself, but involves the whole configuration of apparatuses of observation used in bringing about knowledge of the fetus. In relation to the practices involved in FTPRA, different configurations of apparatuses of observations are mobilised at different moments or phases of FTPRA. In the following I focus on three situations in FTPRA: 1) The visualisation of fetal life in the context of the ultrasound scan 2) the calculation of fetal life in the context of the risk assessment and communication, and 3) decision making about life. On the basis of ethnographic material such as interviews with pregnant women

---

2   To clarify this point Barad develops the concept intra-action to emphasise the sense in which entities emerge through their encounters with each other in a continual process of becoming (Barad 2007: 33). As opposed to the concept interaction, which suggest two entities given in advance that come together and engage in some kind of exchange, she suggest that distinct entities, be it human/nonhuman, the social and the material, do not precede practice. Rather, the components are only distinct in relation to their mutual entanglement (Barad 2007: 33).

and their partners, interviews with professionals and observations of the socio-material practices involved in FTPRA, I frame those situations as constitutive for the making and remaking of boundaries around life, and as such illustrative of a process of bio-objectification, and ask: What kind of phenomena are constituted through these three moments, and how are boundaries around what comes to be constituted as human, nonhuman, subject and object, the normal and the pathological made, contested and acted upon?

Visualising life

The first step on the FTPRA trajectory is the ultrasound scan. During recent decades ultrasound imaging technology has increasingly been used as a routine part of antenatal care in Euro-American countries. The notion of FTPRA as "routine" was also made explicit in conversations with pregnant women and their partners; as one woman said:

> I did not think that much about whether it was a choice or not. It was just like, yeah please, we accept that, just as we accept the midwife and antenatal classes … we take the whole thing, the whole package.

The act of undergoing FTPRA was often explained as an outcome of a fundamental trust in the Danish health care sector and a shared value about using new technology and knowledge in the pursuit of having a healthy baby and a hope of receiving a visual proof that they were expecting a healthy baby.

This was the case for Anita and Allan who accessed and anticipated FTPRA with the hope that the technology would show itself as a step on the road towards having a healthy baby. Anita is 35 years old and she is pregnant with her partner Allan for the first time. Anita and Allan explain how pregnancy until now has been experienced as unreal to them. For the last couple of weeks Anita has felt a bit sick, which she connects to the fact that she is pregnant, but the only other outward sign has been the blue line on the pregnancy test kit. Anita says:

> We have also used it as a proof, that there is something in there … you would like to be confirmed that there actually is something.

When they arrive at the ultrasound clinic a sonographer calls them into the examination room. Anita pulls down her trousers and lays down on an examination couch and the sonographer places the tranducer on Anita's belly. During the scan the sonographer marks out specific points on the image on the screen; she measures the nuchal translucency fold and she points out vital organs such as the heart, the arms, the hands, the bladder and the umbilical cord. During the scan the sonographers states that this one is an active one. Anita elaborates and states that it apparently has inherited Allan's hot temper. Allan giggles and comments that it looks as if it is waving at them. They both smile and Anita sheds a tear. When the scan is over Anita pulls up her trousers and they walk out of the room to await the result. The sonographer goes into another room in order to make the final risk calculation.

During the scan, boundaries are constituted around fetal life through interpretations of what is seen on the image of the screen. Anita and Allan were not immediately able to understand or independently interpret the grey reflected echoes on the screen. However, during the scan the sonographer described the image to them in detail as she pointed out specific body parts and characterised the fetus as active. Anita and Allan elaborated on her interpretation in their conversation. Through this process the image on the screen gradually emerged as an autonomous life with specific characteristics, disconnected from Anita's body. In Baradian terms what Anita and Allan "saw" on the screen can be interpreted as the relational entity of a phenomenon which emerged as an autonomous life through the temporal intra-action of "apparatuses of observation"; the ultrasound scan, Anita's pregnant body, the fetus, the sonographer's interpretation of the image on the screen and their own interpretation of what was said and shown to them. These apparatuses then can be understood as intra-active agents and as such, part of the phenomenon of the child in its emergent state. However, during the ultrasound what was seen was interpreted as what exists. This collapsing of the image of the screen with an ontological stable entity had a strong emotional effect on Allan and Anita. Anita explains:

> We saw how it moved around and bent one of the legs and it waved ... or ... well, it didn't wave (laughing) but it did something with one of the arms. It was such a great emotional experience ... you could see both arms and legs and head with the jaw and everything. It seemed like a human being. It made me feel warm all over. And then it was like, then it became real, you could really see, that this was, well,.it became a real child.

Feminists have argued that the routinisation of ultrasound in antenatal care makes the woman invisible and objectifies her as "maternal environment" (Tyler 2000, Casper 1996, Sandelowski 1994). We might conceive this argument as a version of technological determinism, which holds that technologies "act on their own" and that the use of technologies determines specific kinds of action or meaning. However, the cases in my material contest this image. While, as Petchesky argues, "[ultrasound] images by themselves lack 'objective' meanings" (Petchesky 1987: 78), they need to be translated in practice for them to become meaningful. In Allan and Anita's case this fundamental lack of meaning of ultrasound images gave shape to a space in which interpretations were crafted in the relationship between them, the sonographer and the image on the screen, in order for the image to become meaningful. In this space of action they were not passive or docile. Rather, they participated in a dialogic fashion with the sonographer, in the enactment of the meaning of fetal life. This makes evident the point made by STS scholars that the meanings of technologies

– what they are and what they do – are not a stable and pre-existing practice, but emerge in ongoing ways through the socio-material practices and anticipations involved in their use (Mol 2002, Berg and Mol 1998, Cussins 1998).

As boundaries around fetal life were made during the ultrasound, so were new identities for Allan and Anita. They gradually came to perceive themselves as prospective parents and their relationship to the fetus changed. In their case the process of undergoing ultrasound was also a process through which they were shaped as prospective and responsible parents and they reacted to seeing the image on the screen with great affection. This makes evident that practices of FTPRA not only create boundaries around the human and the nonhuman, but may also involve a reconfiguration of how pregnant women and their partners perceive themselves and their relationship to the fetus.

Calculating life

While the pregnant woman and her partner await the result in the hallway the sonographer will have to calculate an overall risk figure. If the risk figure is assessed as "high" it will open up a space of action and decision making to the pregnant woman and her partner who are offered the possibility of undergoing an invasive test, which is diagnostic, but involves the risk of inducing a miscarriage. The sociologist Nikolas Rose argues in the article *The Politics of Life Itself* (Rose 2001) that in contemporary liberal society, risk calculation has become a major social technology, through which categories of the normal and the abnormal are created. Rose explains:

> ...within the political rationalities that I have termed "advanced liberal" the contemporary relation between the biological life and the individual and the well being of the collective is ... no longer a question of seeking to classify, identify, eliminate or constrain those individuals bearing a defective constitution, or to promote the reproduction of those whose biological characteristics are most desirable, in the name of the overall fitness of the population. Rather, it consists in a variety of strategies that try to identify, treat, manage or administer those individuals, groups or localities where risk is seen to be high. The binary distinctions of normal and pathological, which were central to earlier bio-political analyses, are now organised within these strategies for the government of risk. (Rose 2001: 6-7)

FTPRA is one such strategy through which conceptions of the normal and the pathological are organised. In the following I ask, how are notions of the normal and the abnormal constituted in FTPRA? How are different apparatuses of observations brought about in this process, and what are the implications?

The calculation of a risk figure in FTPRA is a complex matter where several parameters are taken into account, such as the pregnant woman's age-related risk, the size of the nuchal translucency, the crown rump length and the nose-bone

(Nicolaides 2004). After the ultrasound scan has been performed, the sonographer will type these factors into a software programme developed by the Fetal Medicine Foundation, and a final proportional risk figure will be assessed (i.e. 1:45, 1:351 or 1:12000). To demarcate the boundary between a positive (high risk) result and a negative (low risk) result, a cut-off point is of vital importance.

The cut-off point has been contested among Danish hospitals and has ranged between 1:250 and 1:400 (Ekelund et al. 2008). Since January 2007, it was decided to implement a common cut-off point at 1:300, which is in accordance with the recommendations put forward by the Fetal Medicine Foundation. This cut-off point is settled on the basis of complex cost-benefit calculations made with the aim of defining a limit of access to invasive testing that most effectively distributes pregnant women into the categories of high risk (requiring an invasive test) and low risk (not requiring a test). The calculation is based on a large sample of epidemiological data (Nicolaides 2004) and expresses a relationship between detection rate, false positive rates and economic costs. There is a trade-off between these factors, and where the exact limit is set is not objectively evident, but is based on normative and political decisions balancing different interests. As such, the cut-off may be conceived as the result of a translation of a long chain of heterogeneous associations made up of people (epidemiologists, doctors, nurses, politicians, pregnant women), bodily substances (blood, fetuses) and things (computers, software,) and expresses normative, political and economic intentions and values about the desired outcome of the programme on a population level. In this sense the cut-off point may itself be conceived as phenomena emerging in intra-action as a result of a complex configuration of a diverse range of apparatuses of observation.

In observations of the practices involved in the calculations of risk at the ultrasound clinic there was a significant contradiction in the way in which risk was talked about within the room where the calculation was undertaken by sonographers, and outside the room where the risk figure was communicated to the pregnant woman and her partner. Inside the room notions of risk often emerged as very fluid as uncertainties and ambiguities (for instance how to interpret an ultrasound picture) were discussed openly. However, outside the room knowledge about risk was presented to the pregnant woman and her partner in the form of a risk figure, thereby crafted as stable and objective. As such, the FTPRA process was also a process of *purification*, where heterogeneous socio-material practices came to be crafted as objective through its presentation in the form of risk figures (Latour 1993). This process illustrated how, in Theodore Porter's terms, statistics worked as a basic technology for crafting objectivity and stabilising facts (Porter 1995).

Deciding about life

Pregnant women and their partners receiving a risk figure classified as "high" now face the decision about whether or not to undergo an invasive test (amniocentesis or a CVS). Such tests are diagnostic but involve the risk of inducing a miscarriage.

In the guidelines, which formed the basis of the introduction of FTPRA as a routine offer in Danish antenatal care, it was emphasised that choices made on prenatal testing should be made on the basis of non-directive information, so that autonomous and informed choices can be made. The new guidelines on prenatal testing in Denmark invoke a supposed new paradigm of prenatal testing, going from a paradigm of prevention to a paradigm of self-determination (Sundhedsstyrelsen 2004). The guidelines reflect the current general move in the government of reproduction and public health in Euro-American countries to distance itself from previous constraining regulation technologies such as juridical restrictions and directiveness in counselling, towards a greater emphasis on individual decision making and freedom of choice. In his compelling analysis of how bio-power is played out in contemporary neo-liberal society Rose argues that we are witnessing a new form of etho-politics where life is made the object of adjudication. Rose elaborates: "with the state no longer expected to resolve societies' need for health" (Rose 2001: 6) individuals are now governing themselves, taking responsibility for their own health, making decisions about life free from state coercion but in alliance with a plethora of "experts of life" (Rose 2001). Etho-politics is defined as "the self-techniques by which human beings should judge themselves and act upon themselves to make themselves better than they are" (Rose 2001: 18). In this new form of etho-politics, Rose argues, risk has become the organising principle of a "life of prudence, responsibility and choice" (Rose 2001: 18).

Along the same lines the American anthropologist Rayna Rapp suggests viewing pregnant women and their partners making decisions on fetal life in the context of prenatal testing, as "moral pioneers". She states "situated at a frontier of the expanding capacity of prenatal testing" they are "forced to make concrete and embodied decisions about the standards for entry into the human community" (Rapp 2000: 3). Similarly, Rapp invokes a rather individualistic image on how pregnant women and their partners make their way through the complex capacities of prenatal testing and the new spaces of action and decision making emerging from those capacities. I argue that Rose and Rapp come to reproduce an individual image of such trajectories and how decisions are made. Investigating how decisions on prenatal decision are made in Denmark makes visible how prenatal decision making involves much more than individual undertakings. I suggest that decisions may be seen as the outcome of complex processes of meaning making and knowledge production through which boundaries around life (human and nonhuman) and risk (the normal and the abnormal) are made and contested (Schwennesen 2011). This process takes place in the relational space between the institutional and clinical setting and the social life of the individual. In this process, the organisational and clinical settings work as important apparatuses of observation, by constituting some risks as high and others as low, and thus for the decision being made, as I will illustrate below.

In interviews, sonographers explain that they realise the cut-off works to divide pregnancies into good ones and bad ones – a fact which might influence their own professional interpretation of a risk figure as either good or bad. One sonographer says:

> A figure is not just a figure. Working with prenatal testing within this framework, makes me think that when a risk figure is around or above 1:300 then it is not a good situation. It is not nice for me, to go out and give such a result.

If a risk is measured as low according to the cut off, the sonographer will be tempted to consider the situation as good and if a risk is measured as high according to the cut off, she will be tempted to consider the situation as bad.

In cases where the sonographer interprets the result of the risk assessment as high or potentially problematic, she may invite the pregnant woman into a separate room to talk about the risk figure, in contrast to situations where risk is interpreted as "low risk", where the sonographer will give the result of the risk assessment in the hallway. Several sonographers explain that the very act of inviting pregnant women and their partners into a separate room might dramatise the situation and the sense of being in a particular, problematic situation. This is also made evident in interviews where a woman explained to me her experience of being invited into a separate room:

> when she called us into the room I became nervous … I thought that something was really wrong.

As such, the cut-off and the organisation of space work as important apparatuses of observation, contributing to the making of boundaries around risk in clinical practice, which challenge the ideal of non-directiveness and autonomous decision making.

Another significant finding is the way in which the communication of risk figures in itself is experienced as meaningless. A woman receiving a risk figure at 1:668 expresses the sense of insecurity and loss of control which it gives rise to:

> I don't think that I can use a figure which is 1:668 to anything, eh … I know that it is in relation to the general population, but in reality I cannot use that number, because I have nothing to compare with. The only thing I can compare with is the cut of at 1:300 … eh … but it is still rather arbitrary to me, because I do not know whether it is set high or low in comparison to the actual outcome in practice.

One strategy the woman and their partners often use is to ask the sonographer what she would do if she was in their situation. A woman who received a risk figure of 1:164 explained to me her experience of doing so:

> I tried to talk to the sonographer who informed about the risk. What would you do? was my first question, right, what will you recommend me to do, and she said that she could not tell me anything at all. But I felt immediately that I

needed something to compare with. A number is just a number, right. I work with numbers all day, and I know that you can perceive a number from 1000 different views right, and it will mean something different every time.

In this quote the woman expresses the view that risk figures are meaningless if they are presented without any frame or reference points. In their search for frames or interpretative tools the cut-off point often became the primary organising actor, which makes an otherwise meaningless risk figure meaningful. One woman said:

To me, the cut-off point is very important. It is something about letting the experts decide, or something like that, right, because you … well, of cause you have to listen to yourself, but when you are placed in a field where you don't have any experience, then I think I do it, per definition. (woman 40, risk 1:415)

As such, the cut-off worked not only as an interpretative device, shaping decisions in particular ways, but also a helpful ordering tool. As one woman said:

I think it is nice that at least you have a figure you can relate to. I know that it is also rather arbitrary, but then there is at least something you can relate to.

By constituting the cut-off as an important actor in processes of decision-making the individual comes to connect herself to the collective of the population. Hereby, the rationale of prevention implicitly comes to manifest itself as a true and proper way of responding to and acting on risk knowledge. We may understand the cut-off as an *immutable mobile* (Latour 1987) which transports the rationality of effective prevention into the space of possibility that emerges in the context of prenatal decision making and comes to have effects on the ways in which risk is understood and acted upon in the processes of decision making. When the cut-off emerges as a proportional risk figure in the context of prenatal decision making, however, it is detached from the heterogeneous chain of associations from which it is created and thereby crafted as objective and value-free.

When assumptions about autonomy and self-determination are inscribed into the social practice of FTPRA responsibility is transferred only to the pregnant woman and her partner undergoing FTPRA. A woman explains how she experienced the process of having to take on responsibility, on the basis of meaningless risk figures:

Normally I do like to take responsibility and make decisions; I like being responsible and being able to influence things. It fits very well with my personality. But in this situation it was not really a good experience, because I felt like, well, the health professionals are the experts and I know nothing and I was like, well, it is their field and they have the professional knowledge, so it should be possible to get more information or advice. I felt like that … at the same time I knew that I could not demand that. Eh … so I did not really like the

responsibility I had. It was not a decision I felt like facing .at the same time I
knew that I was the only one who was capable of doing that.

What is spelled out here is the way in which FTPRA produces responsible decision
makers by delegating responsibility to its users. This process does not always
consolidate an actual wished-for responsibility, rather pregnant women and their
partners are unwillingly constituted as such. The empirical material illustrates how
apparatuses of observation not only create objects (fetal life) but also subjects
who are shaped in certain ways. In the case of FTPRA pregnant women and their
partners are shaped as actors who must choose, and thereby take responsibility for
the choice which is made.

However, it is possible to challenge the disciplinary effect of the cut-off point.
A woman who received a risk of 1:262 explains to me how she does not see this
risk as "high" which is the reason why she does not want to undergo the invasive
test she was offered. She talks about a past experience she had when she was
pregnant with her first child. She was then offered an amniocentesis (because of
her age) that she accepted. In her conversation with me she recalls how she had
an experience on the way home from the hospital which made her change her
decision on what she would do in the case of a positive diagnosis:

> When I went home from the hospital that day in the bus, a grown up boy with
> Down's syndrome came into the bus and sat with his ticket. It was clear to me
> that he had a life which, eh, which he was able to live and it was not hopeless.
> ... With Down's syndrome you can live about 45-50 years, and there are a lot
> of, sort of help facilities, you will not be hidden away like you would be 100
> years ago.

On the basis of these experiences this woman does not perceive a risk figure of
1:262 as high risk. Her past experiences worked for her as an important apparatus
of observation, through which a possible future with a child with Down's
syndrome is not seen as risky or dangerous. Hereby she mobilises into the present
the somehow marginalised image of a child with Down's syndrome as congruent
with a good life. This illustrates the point that how standards such as risk and cut-
off points are translated in practice, are not a given in the nature of the standard
itself, but are an empirical question, which has to be investigated.

Conclusion

Exploring how fetal life is made to matter in the socio-material practices of
FTPRA through the lens of bio-objectification, makes visible how boundaries
around human/nonhuman, the normal/the pathological and subject and objects
are made through temporal configurations of apparatuses of observation which
are brought about in the production of knowledge about fetal life. In doing so its

basic material/biological form is accepted at the same time as it is considered an effect of an assemblage of relationships made up of material, technological and discursive forms. As such, we are allowed to study fetal life both as a material entity *and* as a process at the same time. Such a framework does not exclude either fetal agency or female agency from the analysis but acknowledges the complex and co-productive relationship between the two. In fact discarding fetal agency from the outset, by creating a universal boundary around who or what has agency, might make invisible the process through which such boundaries are created and how they come to have effects in practice.

The case illustrates how processes of bio-objectification in the context of FTPRA may give shape to new relations of responsibility by attaching people to new entities (of life and risk) and new moral spaces of possible action. This process is neither determined solely by autonomous individuals, nor by technologies, but is made through assemblages of socio-material relationships. If we want to discuss the ethical and social implications of the possibilities which arise from new genetic and reproductive technologies we need to discard predetermined boundaries and entities, and instead consider the ways in which boundaries around what comes to be understood as human and nonhuman, subjects and objects, the normal and the pathological, are made, contested and acted upon in practice and the ways in which they come to have effects in practice.

References

Barad, K. 2007. *Meeting the Universe Halfway: Quantum Physics and the Entanglement of Matter and Meaning*. Durham, NC: Duke University Press.

Barad, K. 2008. Posthumanist performativity: Towards an understanding of how matter comes to matter, in *Material Feminisms*, edited by Alaimo, S. and Hekman, S. Bloomington and Indianapolis: Indiana University Press, 120-54.

Berg, M. and Mol, A. 1998. *Differences in Medicine*. Durham, NC: Duke University Press.

Bohr, N. 1958. *Atomic Physics and Human Knowledge*. New York: John Wiley and Sons.

Butler, J. 1993. *Bodies that Matter: On the Discursive Limits of "Sex"*. New York: Routledge.

Casper, M.J. 1994. At the margins of humanity: Fetal positions in science and medicine. *Science, Technology and Human Values*, 19(3), 307-23.

Casper, M.J. 1996. Reframing and grounding nonhuman agency: "What makes a fetus an agent?". *American Behavioral Scientist*, 37(6), 839-56.

Cussins, C. 1998. Ontological choreography: Agency for women patients in infertility clinics, in *Differences in Medicine*, edited by M. Berg, and A. Mol. Durham, NC: Duke University Press, 166-201.

Ekelund, C.H., Jørgensen, F.S., Petersen, O.B., Sundberg, K. and Tabor, A. 2008. Impact of a new national screening policy for Down's syndrome in Denmark: Population based cohort study. *BMJ*, 337.

Holm, S. 2005. Embryonic stem cell research and the moral status of human embryos, *Ethics, Law and Moral Philosophy of Reproductive Medicine*, 1(1), 63-7.

Holm, S. 2000. Etiske problemer i forbindelse med manipulation af menneskelige fostre. *Origo*, 68, 3-7.

Latour, B. 1987. *Science in Action*. Milton Keynes: Open University Press.

Latour, B. 1993. *We Have Never Been Modern*. Cambridge, MA: Harvard University Press.

Mol, A. 2002. *The Body Multiple: Ontology in Medical Practice*. Durham, NC: Duke University Press.

Nicolaides, K.H. 2004. *The 11-13 Week Scan*. London: Fetal Medicine Foundation

Petchesky, R. 1987. Fetal images: The power of visual culture in the politics of reproduction, in *Reproductive Technologies: Gender, Motherhood and Medicine*, edited by M. Stanworth. Minneapolis: University of Minnesota Press, 57-80.

Porter, T. 1995. *Trust in Numbers: The Pursuit of Objectivity in Science and Public Life*. Princeton, NJ: Princeton University Press.

Rapp, R. 2000. *Testing Women, Testing the Fetus: The Social Impacts of Amniocentesis in America*. New York: Routledge.

Rose, N. 2001. The politics of life itself. *Theory, Culture and Society*, 18(6), 1-30.

Sandelowski, M. 1994. Separate, but Less Unequal: Fetal ultrasonography and the transformation of expectant mother/fatherhood. *Gender and Society*, 8(2), 230-45.

Schwennesen, N. 2011. *Practising Informed Choice: Inquiries into the Redistribution of Life, Risk and Relations of Responsibility in Prenatal Decision Making and Knowledge Production*. PhD thesis, Copenhagen University, Institute of Public Health.

Schwennesen, N., Svendsen, M.N. and Koch, L. 2010. Beyond informed choice: Prenatal risk assessment, decision making and trust. *Clinical Ethics*, 5, 207-16.

Schwennesen, N. and Koch, L. 2009. Calculating and visualizing life: Matters of fact in the context of prenatal risk assessment, in *Contested Categories: Life Science in Society*, edited by S. Bauer and A. Wahlberg. Farnham: Ashgate, 69-87.

Schwennesen, N. and Koch, L. 2012. Representing and intervening: 'doing' good care in first trimester prenatal knowledge production and decision-making. 18th monograph of the *Sociology of Health & Illness*.

Schwennesen, N., Koch, L. and Svendsen, M.N. 2009. Practicing informed choice: Decision making and prenatal risk assessment – The Danish experience, in *Disclosure Dilemmas: Ethics of Genetic Prognosis after the "Right to Know/ Not to Know" Debate*, edited by C. Rehmann-Sutter and H. Müller. Farnham: Ashgate, 191-204.

Sundhedsstyrelsen [Danish Board of Health] 2004. *Nye retningslinjer for fosterdiagnostik* [New guidelines for prenatal testing]. Copenhagen: Sundhedsstyrelsen.

Tabor, A. 2006. Personal communication. Professor in fetal medicine, Rigshospitalet. Copenhagen: Denmark.

Thompson, C. 2005. *Making Parents: The Ontological Choreography of Reproductive Technologies*. Cambridge, MA: MIT Press.

Taylor, J.S. 2000. Of sonograms and baby prams: Prenatal diagnosis, pregnancy and consumption. *Feminist Studies*, 26(2), 391-418.

Tørring, N., Jølving, L.R., Petersen, O.B.B., Holmskov, A., Hertz, J.M. and Uldbjerg, N. 2008. Prænatal diagnostik i Århus og Viborg Amter efter implementering af første trimester-risikovurdering. *Ugeskrift for læger*, 70, 50-54.

PART 3
Generative Relations

The third and concluding part of this book looks at how new forms of life are simultaneously made by and are generative of social, political, economic and philosophical questions about life as an object of inquiry.

Bettina Bock von Wülfingen's chapter examines how techno-scientifically assisted creation of life and related relations of love and loving are intertwined in the question of IVF-embryos in German debates at the turn of the 21st century. Wülfingen argues how IVF embryos are extra-uterine forms of life that come into being through particular imaginative spaces in patterned discursive formations. She shows how the discursive space and the meaning of IVF-embryos constantly moves, depending on the contemporary "German condition" as she puts it. These conditions entail a number of discursive threads – most notably genetics, romantic love and normative sexual relations – framing the debates about IVF embryos. These entwined discursive formations have been generative in that they have given birth to a new bio-object, the figure of a genetized love-embryo, or the "fruit of love" occupying imaginative spaces in the current debates of IVF issues in Germany.

Ingrid Metzler's analysis on the role of states in generating particular conditions of possibility for the emergence and stabilization of bio-objects stresses the point that, despite much ongoing debate about globalization and the erasure of national boundaries, the old state form is still very potent in shaping matters of life. Her analysis focuses on regulation of IVF embryos and related research practices in the UK and Italy. Metzler traces the different legal statuses of the embryos through the last decades and shows how, as a consequence, the whole reproductive fields in the countries have been regulated differently generating also different paths and possibilities for the emergence of embryos in their political, biological and economic forms. These paths, or bio-objectification processes, conditioned by states, as she argues, should be seen performative in their epistemological, ontological and economic senses.

The next chapter, "Growing a Cell *in silico*" written by Niki Vermeulen, examines crucial steps in the creation of a large scale science project attempting at recreating cells in digital form. The project analysed in the chapter takes the paradigm of life-as-information to its extremes in two senses: first, life understood as information after the molecularisation of biology rests on interesting assumptions that detaches actual biological matter from its principles of operation – the "code" that can be cracked with tools and knowledges derived from bioinformatics.

However, and secondly, the idea of decoding the cell in its biological form in order to be recode it on a silicon chip not only changes the idea of cell life but, instead, draws together a set of generative relations that re-articulate the code as a new bio-object within a new field of biology, systems biology. Thus, in this process, the "cell" as an object of life goes through a radical transformation from organic to virtual form. In addition, the process that begins as a local attempt of modeling cells digitally becomes central part of a large-scale biology project, or "big biology" as Vermeulen calls it. This movement generates new relations between academic research and industrial applications, fields of biology, science and policy, and individuals, departments and institutes all around the world.

The application of genetic information in insurance business has been a contested issue in the EU, US and elsewhere. Ine Van Hoyweghen shows in her chapter how new insurance practices made possible by the new biosciences have quite unpredictable generative consequences on social relations. For her, genetic information has the potentiality of assembling new groups together around a particular bio-object ("risk gene") giving rise to new practices of bio-sociality. However, this information also has the power to reconfigure the solidarity between different patient groups. While the use of genetic information in insurance risk calculations for a life has been politically regulated in some countries other types of tests for identifying risks of a living body are still allowed for insurance actuary practices. And, although politically enforced bans for using genetic information have been designed with non-discrimination and "fairness" in mind, Van Hoyweghen argues that this non-discrimination, in fact, might be seen as discriminative in relation to non-genetic patient groups and more traditional tests they are required to take in order to get life insurances. As such, genetic information and 'risk genes' work both as an operator of solidarity between those at risk and of political discrimination for those whose 'condition' are not seen to stem directly from genetic reasons. Moreover, there is no simple solution to this dilemma of social proliferation that emerges together with new bio-objects. Rather, these are generative of new identities and related divisions of the "social", contesting old insurance classifications and reconfiguring solidarity practices according to local circumstances. As such, bio-objects are, as Van Hoyweghen argues, transformative.

The last chapter examines still life as a bio-object. While much of the philosophical tradition in the West has defined life as animation, current practices of cryopreservation have been able to temporarily suspend life without destroying it, keeping its vitality potent. Taking non-human life as subject, Sakari Tamminen carefully analyses the different phases through which frozen-life is created, starting with early modern experiments extending towards more recent research concerning plant and animal reproduction in the twentieth century. Different technologies and practices to manage life are developing, and these processes of bio-objectification, playing with the boundary between life and non-life, have successfully stretched the attribution of life and vitality to inanimate objects. However, at the same time these bio-objects can relatively easily move and circulate around the globe, e.g. making specific species internationally prevalent and thereby reducing biodiversity. But the chapter also argues the other way around as these processes of bio-objectification and the resulting bio-objects are becoming a national concern, restricting the mobility of life and nationalizing bio-objects, making them representations of nationhood and thereby exploiting biodiversity. As a result, Tamminen not only shows the creation of an inanimate, vital bio-object, but also how processes of bio-objectification are both generated by and are generating (inter)national relations, showing deep connections between life, economy, politics and culture.

Sakari Tamminen and Niki Vermeulen

Chapter 9

The Fruit of Love: The German IVF-embryo Turning from Abject into Bio-object

Bettina Bock von Wülfingen

Embryos generated in laboratories do at first sight exemplify the bio-object *par excellence*. They fit the core criteria of our definition: they seem 'out-of-place'-nature, questioning the nature/culture divide as well as the one between 'life' and 'artifact'. As a result of different national regulations they are necessarily[1] embedded in national policies regarding population, especially their status between human and nonhuman, as well as concerning their relation to and that between their parents (which multiply through the laboratory procedures) and technological creators. They are involved in and produce rules for (medical) industry standards, quality measurement practices and quality checkups of their environment, and in their own regard. Entangled in international desires and differences of power, extra uterine embryos are sent around the globe and treated as a commodity. At the same time the 'lab embryo' challenges the realness of the 'lap embryo': nature in the petri dish turns out to be real nature, while uncontrolled nature 'out there', inside dark wombs, becomes unwanted wilderness (Franklin 2006).

All this makes the petri dish embryo a perfect bio-object, one might think at first sight. Yet, it is exactly *not* the being-out-of-place (if there ever was a 'natural' place for an embryo), nor its seeming capacity to challenge the nature/culture- or life/artifact-divide that turns it into a bio-object, that this chapter argues. Such an argument would reenact those modern binaries (in a Latourian sense) and naturalize and universalize the (and any) embryo as bio-object. Instead, although the necessary technology might be in place everywhere, it is very different from country to country and from context to context, when and by what means embryos bio-objectify. Still, officially in some places and in some respect, they seem to resist any (medical, entrepreneurial and research related) treatment and economization. Germany serves as an example in this chapter, being a place where within the past ten years a naturalized and humanized laboratory embryo is questioned more and more in the economic need for it to turn into an (officially recognized) bio-object. This chapter focuses on a particular moment in German history, around the year

---

1 Even if illegally produced, they would not escape these regulations as also their illegal status would determine their 'wild' economic and technological regulation apart from the law.

2000, where for the first time arguments were introduced into public discourse in favour of the manipulation of the embryo that before had been a reprobate or 'abject[2] bio-object'.

When using bio-objects as an analytical tool we want to know how this type of techno-social life comes into being, what its coming into being means and what defines its material and epistemic identity. What turns it into a commodity and ties it to the globalized economy of life? How does its performativity, its agency, change not only technical, legal and economic but (thereby) also social relations?

In this chapter, I argue that it is not only technology that shapes the embryo into a bio-object but rather different discursive measures. Matter and its technological treatment interfere with an economy of 'thinkabilities' (Ariès 1978), closing and opening up 'imaginative spaces' in more Rheinbergerian terms, where the embryo as a bio-object can emerge. If one major characteristic of the bio-object is that it is involved in regulation, than this chapter describes the *regulation of discursive traits* building those discursive formations that enable the disconnect of the embryo from earlier discursive structures (such as 'natural health'), that tended to hold it apart from technology, and to reconnect it to discourses that ease and produce what we call its bio-objectification. The bio-object in this chapter is the 'fruit-of-love-embryo'. It was (is) part of a media hype about new reproductive technologies at a very specific, contingent political moment. In contrast to the non-bio-objectified German laboratory embryo, now different discursive threads, while constituting the space for the 'fruit-of-love-embryo' to exist, at the same time materially re-embedded it in (global) medico-economic contexts.

They combined received anti- and pro-natal discourses: more than just being 'normal', conceiving a child by technological means (instead of 'just so' and without expert-supervision) becomes the essential symbol and intrinsic content of 'real' love, as the parents control the genetic equipment of the potential child. Thus, this type of love (described in more detail below) also resolves an apparent conflict in the German discourse between 'the loss of nature' through IVF or genetic technologies, thus through technological intervention on one hand, and the act that generates life, the latter being tightly connected to the concept of nature, on the other hand. This 'fruit-of-love' can only be valid when having genetic ties to the parents. This fruit is not a fruit of lust, of physical contact, but of a platonic love (project). Love on its part needs to materialize to proove itself. As I will show in the following discussion this bio-object profits from the contradictory existence of modern binaries; of the clash of individualism and love, of nature and culture (as our way to overpower nature), to stabilize itself. It uses and abuses our need to accommodate the frictions between these modern categories, dwelling in their

---

2   The IVF-embryo was an abject bio-object in Germany for most of the time in the sense that it was not meant to exist. It was a 'should-not-be' with a moral connotation. The 'abject', in a butlerian sense, thereby constituted that what was not meant to exist: the 'un-natural' embryo.

betweens and discarding those, which are unfavourable to (neo-)liberal economic processes.

Global bio-objectifications and the German condition

During recent years the regulation of bio-technologies, especially of reproductive and genetic technologies used on human beings, has been a cause of increasing political debates. Individual countries came to very different conclusions: Thus in May 2006, for example, the Danish Parliament decided that public clinics should be obliged to offer IVF even to singles, the full costs being covered by health insurance. In contrast, some years before, the government of Germany, after intense debates of the supposed need for a new (and more permissive) Reproductive Medicine Legislation, decided *against* a change in its Embryo Protection Act (EPA)[3] which is considered to be 'the world's most restrictive law as far as reproductive medicine is concerned' (Beier and Beckmann, 1991). It inhibits any manipulation of a fertilized egg cell after fusion of the nuclei. Since abortion and so-called 'fetus reduction' according to 'medical indication' are allowed, to some the impossibility of screening embryos for deemed serious hereditary diseases and to insert less than the maximum three fertilized egg cells, apparently leading to more abortions after In-Vitro Fertilization (IVF), seemed contradictory (ESHRE 2007). Around 2000 however, Germany faced its first serious challenge to the EPA. The debates about the EPA at that time marked a historic moment.

Up until the mid-1990s ideas about reproductive technologies and the embryo in Germany appeared within two separate fields of discourse, in both popular and scientific literature. On the one hand the discourse of 'overpopulation', and the discourse of 'hereditary diseases' were dominant, on the other hand insufficient reproduction was seen as a problem too, demanding a solution (extinction of the population in post-industrial countries, pension-discourse, infertility as disease and affliction). Thus anti-natal and pro-natal topics appeared as separate issues. Discourses of reproductive *medicine* revolved mainly around the individual, such as prevention of 'genetic diseases' in the unborn (thus stressing genetic risks; Rose 2006, Novas and Rose 2000, Giddens 1991), and assistance for heterosexual couples suffering from infertility. Meanwhile any treatment *of the embryo* to solve these individual problems and interests beyond e.g. for research purposes, was legally impossible.

What happened in 1999 was that after many years of Christian-conservative rule, the 1998 newly-elected social-democratic chancellor Gerhard Schröder promised to turn Germany into a location for biotech-business. The IVF-embryo of course plays a major role in this: new reproductive and genetic technologies make

---

3   My acronym; an exception being the reform introduced by the law regulating the import of stem cells.

hitherto unavailable 'human nature' technologically accessible and productive. Clarke (1998) describes the 'newness' of this process regarding reproductive and genetic technologies as a departure from the interest in control (of physical processes, population) towards the search for economic niches concerning the now available material for bio-chemical, statistical or pharmaceutical procedures. Embryos generated in laboratories – be it for reproductive and not research purposes – are resources entangled in global bio-economic networks (Waldby 2008; and Brown, Chapter 4 in this volume).

Apart from many other actions often accompanied by public critique, Schröder installed a new ethics commission parallel to the one already existing in parliament, and the federal and national ministers of health invited the public to many hearings about the EPA. The media followed and triggered these political debates with great interest, often inviting international experts[4] to present their visions of the future utilization of reproductive and genetic technologies, or to comment on the way other countries were already making use of these technologies. It was in this complex situation, in which the embryo was located between received discourses depicting conception and reproduction as necessarily 'natural' and intimate processes that oppose a connection to technology and economy on one hand, and a sudden atmosphere of bio-technological 'enlightenment' on the other, that a new, specifically German bio-object was produced: the geneticized 'embryo-of-love'.

Below, I present results from an analysis of the usually enthusiastic and utopian media contributions in the quality press in Germany from 1995–2003. I focus on the new architecture of Foucauldian objects which changes imaginative spaces in the field of reproduction in a way that allowed this bio-object to come into being. The chapter thereby reconfigures the bio-object of '*the* German embryo', i.e. the one known so far as being enacted through the EPA. The latter, similar to the 'Italian embryo' (see the chapter by Metzler in this volume), had rather been an abject bio-object: by legal means it tended to resist bio-objectification, by being naturalized (together with its relationship to the parents). In contrast, the articles in the German quality press discursively established laboratory technologies as the 'normal' way of 'making' children.

In what follows I will summarize the most relevant results.[5] I will then further interpret them by mapping the central objects whose relational changes shape out and co-construct the fruit-of-love as a bio-object. I will first explain my method of discourse analysis connected to the bio-object as analytical tool and then proceed to demonstrate the emergence of the genetic fruit of love and how discursive elements newly combine common objects such as love, sex and gender, autonomy and health. Finally, these objects will be integrated into

---

4   So labelled explicitly by the respective magazine or newspaper.

5   More quotes exemplifying many of the statements, as well as results here referred to are presented in much more detail Bock von Wülfingen 2007. In order to deal with the space limits of this chapter I chose just some of those which support the major arguments in this chapter.

surrounding and preceding discursive threads, which in turn discursively 'feed' the bio-objectification of conception and result in the new combination of the genetic fruit of love.

Discursive threads, objects and imaginative spaces

In contrast to Anglo-Saxon discourse analysis related to linguistics, the French discourse analysis referred to here draws on Michel Foucault's concept of discursive formations (Foucault 1972, 1991). A distinguishing mark of discourses and the statements they comprise is that they, in contrast to non-discursive utterances, are being articulated from *institutionalized* locations (Foucault 1972: 50-51), such as a clinic, a laboratory or a library. Thus it is not individual speakers, but institutions, systems, pedagogic norms and legal conditions which grant a right to produce and apply knowledge (ibid.). The speakers, however, become significant to a discourse once they carry a specific designation: in the present texts, they are presented as experts of genetics or reproductive medicine, and always in conjunction with an academic background – 'professor in'. Thus, according to Foucault, they serve to legitimate the relevance of discursive incidents (Foucault 1969). Within the texts at hand the institutions relevant to the analysed discourse are on the one hand the academy (or clinic) and on the other hand those 'quality press' journals, which adhere to the conservative journalistic ethos of objective reporting. They mark the respective contributions (essays or interviews) explicitly as 'original' statements by specialized experts.

As this investigation is quite limited in terms of the space and time that the material embraces, I do not talk about *a* discourse but about discursive threads, which are nonetheless part of a discourse. The respective discourse they are helping to construct could be called 'human bonding is a question of matter' – produced within the postmodern capitalization of affect (Boltanski and Chiapello 2007).

The bio-object is, in this chapter, both an epistemic tool and a Foucauldian object. It is a Foucauldian object in the sense that it is produced by discourse and institutions, which at the same time are produced and shaped by the bio-object. The Foucauldian episteme and that of the bio-object show useful overlaps: The bio-object opens up epistemic spaces and helps to investigate which epistemic spaces need to open up to enable the specific bio-object to come into being.

To also more intensively conduct a Foucauldian micro-analysis I selected those articles which argued in favour of a relaxation of the regulation of new reproductive and genetic technologies in Germany. They all purported to be biomedical 'expert statements'. Thus 38 articles were examined for their metaphors, act schemes,[6] discursive threads and the architecture of objects. My macro-analysis comprises a variety of material, such as science-fiction books by experts on the future of human reproduction and reports by political committees on new reproductive and

---

6   See for a critical discussion of these methods Bock von Wülfingen 2007.

genetic technologies (both from Germany, and, for the sake of contrast, also from the US). The majority of this material is taken from those German newspapers and journals which have the greatest print run, are deemed 'quality press' and are used as resources both by politicians and the media (such as the *Frankfurter Rundschau*, *Spiegel* etc.; Informationsgemeinschaft zur Verbreitung von Werbeträgern, 2000). The material analysed thus comprises about 1,000 German media contributions published between 1995 and 2003.[7]

Major discursive threads in the bio-objectification of German reproduction

The conventional medical discursive thread in Germany concerns itself with the restoration of physical functioning, defined as the functioning of the human body according to its natural limits (e.g. Hucklenbroich 2004; Lanzerath 2004). In the federal republic, conflicts about the introduction of new technologies such as genetic testing had been prevalent far into the 1990s consisting of an opposition of nature (which was to be conserved in the way it was imagined to be when untouched by humans) and technology (that would interfere with nature and was, in any case of doubt, risky) (Bock von Wülfingen 2010). Thus, in-vitro-fertilization was also only accepted in as much as it served nature (e.g. Bundesministerium 2004). This includes the situation that, by order of the German Medical Council, only heterosexual (at best married) couples may be offered IVF.

Many of the articles analysed refer to this value system, but others introduce a 'post-sex/gender' thread, which grants all people the right to access technological aid for reproduction regardless of their physical condition, departing from the dominant discourse. In my analysis, I found an intricate network of about 20 different discursive threads, which were often contradicting and overlapping, yet always dependent on one another. They could all be mapped in scenarios of individual autonomy on one hand and heteronomy on the other hand: autonomy especially referred to the opposition to a society which restricts this technology (e.g. Silver, in Petermann and Paul 1998); but also to the opposition to natural limitations, such as the genomic or natural 'game of dice' (Reich 2000: 206, Djerassi 1999: 51) in uncontrolled conception. Heteronomy was the feature of scenarios of market-determinism, which claim that certain technologies would prevail in the long run since this 'progress runs us over' (Stock, in *Süddeutsche Zeitung* 1998) and is 'unstoppable' (ibid.) anyway; or scenarios of biological determination, like those referring to the one-way relationship between one 'gene' and its suggested phenotypical expression.

---

7   The time span of 1995-2003 for the analysis was chosen after Graumann (2002), who dates the beginning of the ethical debate about biotechnological human reproduction, especially about cloning, with the birth of the cloned sheep Dolly in 1997. As a matter of caution, I also looked at the two preceding years. In fact, the present discursive thread started in 1996.

Within this framework the apparently radical concept of laboratory conception as the standard way to reproduce appeared for the first time in public statements by international experts in German media in 1996. It amalgamated other discursive threads, such as 'conception beyond sex and gender' and 'health-promising genes'. Some quotes exemplify this: Heterosexuality would be unnecessary for reproduction and 'some women will welcome this' (Silver 1998: 145), with 'each carrying the clone of her [female] partner' (Green 1999: 64). 'Since reproduction is secured [by the new reproductive and genetic technologies], there is no reason why men should not block their seminal ducts and women their Fallopian tubes' (Baker, in *Focus* 1999: 163), so that accidental pregnancies without expert intervention become impossible. Then 'we will enrich the free spaces on a new chromosome with artificial genes' (Stock, in *Süddeutsche Zeitung* 1998), and 'traditional reproduction will gradually vanish' (ibid.). Such concepts, which digress from heterosexuality as the singular 'healthy' precondition for biological reproduction, had been unthinkable until the mid-1990s, when bio-evolutionary assumptions about the species-typical functioning of the human body prevailed.[8] Thus these texts recombined categories (or 'objects' in Foucauldian terminology) to create these new imaginative spaces (see Foucault 1972) while bio-objectifying reproduction.

The more recent discursive threads, appearing in texts from 1998 onwards, were dominated by a shift, viewing the application of reprogenetics less in terms of healing, but increasingly in terms of liberation from constraints and of enabling. The scenarios of autonomy often referred to 'equal rights', which roughly meant that they advocated emancipatory concerns, for example for women, older people or so-called homosexuals. However, each of the references to equal rights is characterized by an elliptic-deterministic argumentation: without substantial justification, the reasons for social success, for the limitation of action or for discrimination (e.g. for a certain skin colour) are assumed to be biological (thus related to the skin of the person discriminated against). If a form of discrimination could be viewed as being socially determined, its cause is found in biological differences – and biological solutions are offered: 'So far, the problem of reconciling a career with the desire for children has demanded legal or political solutions [in terms of equality for mothers at the workplace]' but now medicine would solve this problem once and for all. 'If a patient wanted to freeze her egg cells at the age of 30 to have them re-implanted at the age of 40, it would be her own decision' (van der Ven, in Lakotta 2001: 186). This would lead to a shift within 'the power-relations of man and woman' (Djerassi, in Thimm and Traufetter 2000: 210).

Accordingly, the phenomena of disease and health are generally viewed as genetically determined: new reproductive and genetic technologies are seen as applicable within four broad areas, which point to the understanding of what needs to be 'cured' or enhanced. Firstly, the area of application suggested in the texts broadens its scope from the idea of a mere healing to the idea of genetic prevention

---

8   For more detail, see Bock von Wülfingen 2002.

(from 1998 onwards). Thus 'genetic technology should provide every citizen with a basic equipment [...] similar to a protective vaccination' (Silver 2000: 147). The next step, mainly taken in later texts after 2000, views the aim in the promotion of well-being and life chances, which is similar to conventional definitions of health from the 1980s, such as the one applied by the World Health Organization. This understanding of health offers a basis for the idea that 'genetic health' is a precondition for autonomy and the capacity to act, which are seen as essential advantages in social competition: 'Parents have a strong desire [...] to guarantee their children a healthy and successful life' (Silver 2000: 146), by 'providing them with good genes', like those which 'influence brain capacities' (Silver 1998: 144).

Consequently, equal rights arguments would allow for a distribution of such medical resources befitting a welfare state. As several authors in the material analysed stress, this would also be valid for genetic optimization: 'In many European countries children disregarding of the wealth of their parents receive a more or less similar health service and education. Under these conditions genetic manipulation, which only rich parents can afford, would be deemed immoral, as unfair against those children that were excluded' (Silver 2000). This would mean 'access for all – genetic technology on sickness certificate' (Stock 2000).

Finally, there is the identity-related demand that 'people should be free to choose which characteristics they want for their children' (Baker, in *Focus* 1999: 163).

In this way, these post-sex/gender (or de-sexualized) and health-related scenarios of conception frequently refer back to widely held concepts of genetic determinism. New in these ideas is the departure from the received view which reflected the conventional medical concepts emphasising the healing aspect, which sees reproductive technology exclusively as a service to assist the naturally given functioning of countersexual couples. It is at this point of departure 'from natural' health where the epistemic space for such a hitherto illegitimate use of technology needs to be produced. One discursive threat to its production was described above; the equal rights-argument. A much stronger one is the new bio-object, the love-embryo.

The appearance of the geneticized love-embryo

The wish to have a child and the desire for reproduction usually play a major role in explaining the use of IVF. The discursive threads which I examined no longer relate the wish to have a child to the category of a universal human nature, but relate it to a different object, independent of physical functions and nature but still referring to a modern value: intimate love within a couple and for the child. As this child needs to be genetically related, I call it in this context an own-genes-child (commonly referred to as 'biological child' [*sic*]). Where before the healing-orientated reproductive technology was legitimated by such categories as risk and responsibility, now love justifies all departures from the conventional concept of health and disease grounded in physiology and evolution. The new architecture of these objects – love, responsibility and entitlement – is co-constructed in a way that inevitably results in the application of new reproductive and genetic technologies: It is assumed that two people in love, regardless of their physical constitution, will inevitably develop a desire to have a child and, if this is not realizable with an own-genes-child, will suffer accordingly. As will be shown below, love now creates a connection between the parents' responsibility for their future child and new reproductive and genetic technologies; and love also explains to society their right to access these technologies, because 'the only problem they [two women] have, is that they cannot conceive with their loved one' (Dahl 1999: 312). If, due to their mutual infertility, it is 'the only possibility to make a desperate couple happy', even cloning or in-vitro-fertilization at a relatively advanced age, should be allowed. Thus 'children conceived in a laboratory are always wanted babies. They are better loved than those conceived naturally' (Djerassi, in Blech and Trauffetter 2002: 77). Within the scenarios of autonomy, the love for the child socially entitles everybody to be given access to relevant technologies, (implicitly) also singles: the (cloned) 'child and its mother, being identical twins at the same time, will be of one heart in the truest sense' and this is why 'people, who would otherwise have no genetic relation to their child, need cloning technologies' (Green 1999: 65). 'Every couple has the right to decide, when and how they want to have a child' (Katzorke, in Paetsch 2003: 149), and thus care has to be taken that legal restrictions of relevant technologies do not 'circumcise the right to biological offspring' (Silver 1998: 142). In the macro-discourse analysed, the publicly discussed search by an American deaf, lesbian couple for a donor who would provide a 50 per cent chance of generating an equally deaf second child, was commented on approvingly: 'The most important issue is that children receive love, a sense of security and support. [...] After all, everyone is selective regarding the partner with whom they want to create children' (Berndt 2002). In this way, the categories which have before been excluded by a physiological and evolutionary understanding of healthy, thus natural and non-technological reproduction – love, equal rights (regarding access to new reproductive and genetic technologies) and identity – are interwoven, integrating those bodies into the discourse. In contrast to the conventional concept of a 'natural' desire to have a child, which could only be imagined as an expression of physiologically and evolutionary healthy bodies, the parental bodies now become irrelevant. The emphasis instead lies on the genetic condition of the projected child, the fruit-of-love, which, however, does not need to conform to 'universal' ideas of health either, but can be created according to diverse parental demands, even if those contradict such former values as best natural human health.[9]

This analysis of Foucauldian objects reveals that this type of geneticization of love and of its fruit works by introducing the 'life chances' discourse into

---

9   For an overview, see also the diagrams of the previous and the present concept of the desire to have a child in Bock von Wülfingen 2007.

reproduction. It reshapes 'risk' as a link between identity and self-sufficient autonomy and thereby renders the tight connection between intimate love and the geneticized embryo plausible. This actualizes the analysis of the 'somatic individual' by Novas and Rose (2000): They describe the change in genetic counselling over the past 30 years. Instead of only addressing genetic risk as a threat to the health of the individual the discourse changed towards promoting 'life chances' and the 'quality of life' of the respective individual. The German texts analysed not only address the enhancement of happiness and wellbeing of the child as an issue of genetic intervention in the name of love but also the identity and creativity of the prospective parents. The 'genetic self' thereby extends into the next generation.

Love and identity

According to Luhmann, the rise of intimate love is closely linked with the concept of the autonomous subject, thus being evocative of connections with 'genetic health'. Intimate love is mainly perceived as a process of negotiating identities, based on confessions (both in a religious and judicial sense), on inclusion and exclusion. It serves to fill the gap of 'desire for intimacy' (Luhmann 1998), left by increasingly improbable identities, adhering to modernity's demand for individuality and flexibility. Correspondingly, intimate love lends itself as a suitable glue connecting categories and people in the introduction of new reproductive and genetic technologies.

Love and identity are also closely tied in to the scenarios discussed above concerning the desire to have a child. The identity of the parents is of increasing relevance to new reproductive and genetic technologies, often replacing former discursive threads about responsibility and risk. The material examined suggests that, with the help of genetics, parents have a free choice of qualities for their future child, thus addressing the demands on identity of those parents-to-be. In fact this fits with findings according to which parents who are faced with having to choose donated sperm or egg cells show highly diversified interests, concerned more with their own identity and the 'passing' of traits to the child than with criteria conforming to a general standard. Instead they adhere to criteria set by their immediate environment, which might well diverge from the norm (Beckera et al. 2005).[10]

---

10 Although these criteria seem to be defined by the overall concept of cognate 'similarity', the results gained by my macro- and micro-analysis call for a closer discussion of the different biological, historical, cultural-theoretical or governmental concepts of eugenics in relation to these shifts (some criteria are offered in Bock von Wülfingen 2007). In accordance with Clarke's discussion of the 'new' in new reproductive technologies as targeting pharmaceutical niche markets rather than aiming at control (1998), Rabinow and Rose (2006) decidedly call this type of biopolitics "capitalism and liberalism, not eugenics' (ibid. 211).

The discursive threads examined in this analysis mainly refer to *couples*. They are supportive of 'homosexual' identity of same-sex couples and their sexuality appears to be acknowledged, since assistance is offered for their desire for reproduction. However, only with the creation of the (planned) own-genes-child, the genetic love-embryo, is intimate love re-integrated into the material world, and homosexuality 'really' normalized.

The threads of discourse analysed relating to identity and emancipation profit, in every respect, from the former legal and political application of an evolutionary-naturalist concept of reproductive health. It pathologized and excluded bodies and thus participated in creating some of those identities in the first place. Thus – in a very Foucauldian sense of the mutual dependence of dominant and 'emancipatory' discourses – the discursive matrix of the new reproductive and genetic technologies as presented in the texts analysed here offer solutions to the historic naturalization of exclusion in the form of a geneticized inclusion.

This chapter described the coming into being of an epistemic space in which the bio-object, a geneticized 'love-embryo' could appear. This bio-object is entwined with global bio-economies and at the same time bound to the specific German context of the debates about a new law on reproductive medicine that would replace the (deemed restrictive) EPA. The geneticized love-embryo in its turn was entangled in attempts at commenting on – if not changing – the German situation. It might (have) come into being wherever the embryo resists bio-objectification and where at the same time a modern value system as described above is in place. I have shown in this chapter, that this bio-object is not only a 'taxonomy hooligan'[11] (a characteristic of bio-objects stated elsewhere in this anthology) but also a value-rowdy. It comes into being by playing with traditional values and by producing and living in their betweens. It conjures some to stabilize itself – such as naturalism in the specific sense of genetic-determinism, autonomy, equality, individualism, intimate love and autonomy – and scrunches others – such as 'untouched' nature, 'natural' bodies and heterosexuality.

References

Ariès, P. 1978. La contreception autrefois. *L'histoire*, May/June, 36-44.

Beckera, G., Butler, A. and Nachtigall, R.D. 2005. Resemblance talk: A challenge for parents whose children were conceived with donor gametes in the US. *Social Science & Medicine*, 61, 1300-309.

Beier, H.A. and Beckmann, J.O. 1991. Implications and consequences of the German Embryo Protection Act. *Human Reproduction*, 6, 607-8.

Blech, J. and Traufetter, G. 2002: 'Laborbabys werden mehr geliebt'. *Spiegel*, 4, 76-7.

---

11   We owe this term to Lena Erickson, another contributor to this book.

Bock von Wülfingen, B. 2007. *Genetisierung der Zeugung: Eine Diskurs- und Metaphernanalyse reproduktionsgenetischer Zukünfte.* Bielefeld: Transcript.

Bock von Wülfingen, B. 2010. Human genetic technologies in international contexts: Local ideas, cultures and concerns in comparison, in *Ethik transdisziplinär: Genetic Screening,* edited by M. Fischer and M. Hengstschläger. Salzburg and Frankfurt am Main: Peter Lang, 367-84.

Boltanski, L. and Chiapello, E. 2007. *The New Spirit of Capitalism.* London, New York: Verso.

Bundesministerium für Gesundheit und Soziale Sicherung 2004. Bekanntmachung des Bundesausschusses der Ärzte und Krankenkassen über einen Beschluss zu Änderung der Richtlinien über ärztliche Maßnahmen zur künstlichen Befruchtung (Richtlinien über künstliche Befruchtung). *Bundesanzeiger,* 13(21 January 2004), 910.

Berndt, C. 2002. Der Wunsch nach dem fehlenden Sinn. *Süddeutsche Zeitung* 93.

Clarke, A. 1998. *Disciplining Reproduction: Modernity, American Life Sciences, and 'the Problems of Sex'.* Berkeley, CA and Los Angeles: University of California Press.

Dahl, E. 1999. Sollten lesbische Paare Zugang zur künstlichen Befruchtung haben? *Ethica,* 7, 307-13.

Djerassi, C. 1999: Der entmachtete Mann. *EMMA,* 5, 50-51.

ESHRE (European Society for Human Reproduction and Embryology) 2007. Germany's embryo protection law is 'killing embryos rather than protecting them'. *Public Release ESHRE,* 4 July 2007 [online]. Available at: www.eurekalert.org/ pub_releases/2007-07/esfh-gep070307.php [accessed 10 July 2010].

Focus 1999. Der Mensch wird seine Reproduktion bald voll steuern (interview). *Focus,* 22, 163.

Foucault, M. 1969. Qu'est-ce qu'un auteur? *Bulletin de la Société française de philosophie,* 63, 73-104.

Foucault, M. 1972. *The Archaeology of Knowledge & the Discourse on Language.* New York: Pantheon.

Foucault, M. 1991. Governmentality, in *The Foucault Effect: Studies in Governmentality,* edited by G. Burchell, C. Gordon and Peter Miller. Hemel Hempstead: Harvester Wheatsheaf, 87-104.

Franklin, S. 2006. The cyborg embryo: Our path to transbiology. *Theory, Culture & Society,* 23(7-8), 167-87.

Giddens, A. 1991. *The Consequences of Modernity.* Cambridge: Polity Press.

Graumann, S. 2002. Die Rolle der Medien in der Debatte um die Biomedizin, in *Kulturelle Aspekte der Biomedizin: Bioethik, Religionen und Alltagsperspektive,* edited by S. Schicktanz, C. Tannert and P. Wiedemann. Frankfurt am Main and New York: Campus, 212-43.

Green, R.M. 1999. Mein Kind ist mein Zwilling. *Spektrum Spezial,* 4, 62-5.

Informationsgemeinschaft zur Verbreitung von Werbeträgern 2000: IVW Praxis, Bonn: IVW.

Lakotta, B. 2001. Kind in der Warteschleife (interview). *Spiegel*, 4, 186.

Lanzerath, D. 2004. Von der Heilbehandlung zur Anthropotechnik. Krankheit als normatives Konzept. Talk paper, in *Nationaler Ethikrat: Wortprotokoll. Niederschrift über das Forum Bioethik*, 14 April 2004, Berlin.

Paetsch, M. 2003. Brauchen wir neue Gesetze für die Fortpflanzungs-Medizin? (interview). *Geo 8*, 149.

Luhmann, N. 1998. *Love as Passion: The Codification of Intimacy*. Stanford: Stanford University Press.

Novas, C. and Rose, N. 2000. Genetic risk and the birth of the somatic individual. *Economy and Society*, 29(4), 485-513.

Petermann, J. and Paul, R. 1998. Gefährlicher als die Bombe (interview). *Spiegel* 29, 142-5.

Rabinow, P. and Rose, N. 2006. Biopower today. *BioSocieties*, 1(2), 195-217.

Reich, J. 2000: Erotik in der Cyberwelt. *Spiegel*, 48, 204-6.

Rose, N. 2006: *Politics of Life Itself: Biomedicine, Power and Subjectivity in the Twenty-first Century*. Princeton, NJ: Princeton University Press.

Silver, L. 2000. Eingriff in die Keimbahn. *Spiegel*, 1, 146-47.

Stock, G. 2000. Der Geist aus der Flasche. *Spiegel*, 15, 190-92.

Süddeutsche Zeitung 1998: Klon der Angst (interview). *Süddeutsche Zeitung*, 11 April 1998.

Thimm, K. and Traufetter, G. 2000. Küss die Hand, gnädiges Ei (interview). *Spiegel*, 48, 210-12.

Waldby, C. 2008. 'Oocyte markets: women's reproductive labour in embryonic stem cell research' *New Genetics and Society*, 27(1), 19-31.

World Health Organization 1969. Genetic counselling. Third report of the WHO expert committee on human genetics. *Technical Report Series* 416. Geneva: World Health Organization.

World Health Organization 1986. *Ottawa-Charta for Health Promotion*. Geneva: World Health Organization.

Young, I.M. 1987. Impartiality and the civic public: Some implications of feminist critiques of moral and political theory, in *Feminism as Critique: Essays on the Politics of Gender in Late-Capitalist Societies*, edited by S. Benhabib and D. Cornell. Oxford: Blackwell, 57-76.

Chapter 10

On Why States Still Matter: In vitro Fertilization Embryos between Laboratories and State Authorities in Italy

Ingrid Metzler

In this chapter, I seek to reflect on the involvement of states in the making and stabilization of vital phenomena as bio-objects.[1] My central point is that states do (still) matter.

This is meant to imply, first, that states are still important and that some of us might have been rash in announcing their demise. Of course, not all of us were ready to dismiss the anabiotic Leviathan. Scholars such as Sheila Jasanoff have taken pains to show that despite the intensification of movements of subjects and objects across formerly firmly entrenched and now leaky national frontiers, such boundaries continue to make a difference in the contemporary geography of the bio-sciences and should therefore not be eliminated from our research agenda (Jasanoff 2005a, 2005b). Yet others have started to turn to political authorities and spaces of different scales, exploring politics on a 'global' (Prainsack and Naue 2006) or 'subpolitical' (Beck 1993, 1997, Vries 2007) scale, implying that, now, it is in these spaces that 'real' politics—i.e., the politics that matters—takes place. I do not want to dismiss these 'new political spaces' (Hajer 2003) as unimportant. Yet in this chapter, I argue that their importance makes neither state authorities nor, loosely borrowing from Paul Rabinow's terms (1999: 7), 'political spaces as we know them', meaningless. States do still matter—*next to* these new political realities. They matter because they matter in yet another, a second, sense (Law

1 This chapter draws on research conducted in the context of the research project PAGANINI (Participatory Governance and Institutional Innovation), funded under the Sixth EU Framework Programme, Contract no. CIT2-CT-2004-505791, available at: http://www. paganini-project.net. Its writing has been enabled by the project 'Biomarkers: Towards the Governance of an Emerging Medical Technology', funded by the Austrian Genome Program GEN-AU. I am grateful to my colleagues at the Life-Science-Governance Research Platform at the University of Vienna, in particular to my supervisor Herbert Gottweis for his continuous support and encouragement, and to Christian Haddad and Georg Lauss. The continuous exchange with them makes my work not only slightly more productive; it also makes it a lot of fun. I am also grateful to the editors, for their comments and for their patience, and to my Italian interviewees who made this chapter possible.

2004, Moser 2008): they *do* matter, i.e., they enact or—at least—help enact matter, such as subjects and objects, or—indeed—bio-objects. Yet how do states help enact bio-objects? And what sort of states help enact bio-objects? In this chapter, I seek to start to reflect on these questions.

I argue that (some) states help make bio-objects (among others) by enacting laws and regulations and by entrusting authorities with the implementation of these laws and regulations.[2] I understand these laws and regulations not as things that come after the bio-object. Rather, I suggest reading them as part of those heterogeneous assemblages that *may* stabilize vital phenomena as bio-objects. Hence, I interpret governance—in a quick and 'sloppy' definition as the enacting and implementing of laws and regulations by (state) authorities[3]—not as a layer that is added onto pre-existing stable objects; nor do I think that governance is triggered by the proliferation of bio-objects or a mere *re*-action to their emergence. Rather, I suggest interpreting laws and regulations as generative or 'performative' (Lezaun 2006) in the sense that they may help purify ambiguous phenomena and stabilize them as bio-objects. Such laws and regulations might well pre-exist the making of laboratory entities. They might come before or after such *entities*. But they do not come after the bio-*object*. In my (admittedly very narrow) use of this term, there is no neat bio-*object*, or bio-Object, without regulation.

I seek to make this argument more robust, by exploring moments of the public life of Italian in vitro fertilization (IVF) embryos. It is here that this story becomes rather complicated.

The first complication derives from the 'embryo part' of this story. Indeed, IVF embryos are ambiguous entities that are hardly ever straightforwardly bio-objectified in practice (see also Bock von Wülfingen's chapter in this volume). IVF embryos exemplify the importance of the hyphen in between the 'bio' and the 'object' in the sense that the more we learn about a particular phenomenon of life (or 'bio') by making it an object of our knowledge practices, the more we might question such practices or expect a plethora of regulations that enshrine conditions under which such 'bios' can be objectified. IVF embryos trouble a resilient dichotomy that is central to the ways in which Western societies are governed: the dichotomy between 'subjects' and 'objects' (Hoeyer et al. 2009)—not least because embryos and fetuses have been continuously made objects of knowledge

---

2   The statement that *some* states help make bio-objects is meant to emphasize that not all states do. I will return to this point in the end of this chapter.

3   This definition is quick because it is merely meant to facilitate the reading of this chapter. It is sloppy as it zeroes in on laws and regulations, which are—no doubt—by far not the only means through which subjects and objects are governed in practice (Barry 2001, Law and Mol 2008). Moreover, it also fails to consider that such laws and regulations might well come into being through interactions that transcend the boundaries of state authorities (Hajer and Wagenaar 2003, Sørensen and Torfing 2007). However, *for the moment* I am less concerned with how these laws and regulations are done; rather, I am more concerned with what these do.

practices over the past two centuries, generating knowledge that made them move ever closer to living human subjects (Morgan 2009). I am by no means suggesting that IVF embryos and embodied embryos and fetuses that dwell in women's wombs, or aborted fetuses that have been removed from women's bodies, are politically, morally, and ontologically equivalent. In contrast, as I show below, the politics of categorization of IVF embryos, involves complicated struggles of entanglement and disentanglement between embryos 'out of place' and fetuses in their 'natural space' (Douglas 1994 [1966]). Yet, they are often actively related to one another. Hence, not least because embryologists had carefully collected and exchanged the flesh of aborted embryos and fetuses in the past (Hopwood 1999, 2007, Morgan 2006, 2009), and also thanks to all the efforts from the latter half of the 20th century to visualize embryos and fetuses in women's wombs (Petechsky 1987, Schwennesen's chapter in this volume), IVF embryos could neither be self-evidently classified as a cluster of cells nor could they easily be categorized as subjects or—indeed—persons. They were sitting uneasily in between these categories; fitting into both and neither at the same time, these entities qualified as ambiguous s/objects, or—using Klaus Hoeyer's (2010) term—as 'ubjects'.

To be sure, not all regulatory debates in which IVF embryos were entangled in over the past three decades were characterized by a collective sense-making on whether to categorize IVF embryos as 'subjects', i.e., as forms of human life that should be included into the political community, or as 'objects', i.e., as forms of bare human vitality. However, such a politics—or, indeed, *P*olitics (Agamben 1998)—of categorization (and of inclusion and exclusion) was sometimes barely concealed under the surface. In many cases, the legal responses that national collectives ended up with enshrined complex sets of boundaries that washed away a little bit of the ambiguity of these 'ubjects', for instance by stabilizing a neatly defined part of them—perhaps—not necessarily as (tradeable) 'bio-objects' but—at least—as 'research objects'.[4] However, in February 2004, Italy enacted a law that refrained from drawing such boundaries around early forms of human life, categorizing all IVF embryos as 'involved subjects' whose rights would be protected by the very 'norms in matter of medically assisted procreation', which is known as Law 40 (Repubblica Italiana 2004). It is from this Italian part of the story that a second complication arises.

Indeed, whilst IVF embryos are hardly ever smoothly bio-objectified, Italian IVF embryos in particular do not quite fit into this vocabulary. In contrast, many of the activities of Italian policy makers can be read as attempts to inhibit a bio-objectification of IVF embryos, to an extent that rather than as bio-objects Italian

---

4   I am making a difference between 'research objects' and 'bio-objects', as in my (admittedly) narrow understanding of this term not every 'research object' amounts to a 'bio-objects' (whilst every 'bio-object' is also a 'research object', yet a research object that starts to have a life of its own, travelling beyond the boundaries of laboratories). In this chapter, then, I am mainly interested in developing an understanding of the ways in which regulations and laws are entangled in the web of relations that make up bio-objects.

IVF embryos seem to qualify as 'citizen subjects' (Filippini 2004, Hanafin 2007, Metzler 2007, 2011). In other words, I chose to exemplify the claim that states, their authorities and practices help enact bio-objects by discussing a case in which state authorities did their very best to impede such a process. I am aware that it is awkward to argue for the need to take into account a presence by exploring an absence. However, I think that it is still helpful to explore a case in which an entity is *not* stabilized as a bio-object in order to develop a better understanding of the uneven and contingent geographies in which such objects dwell. Zeroing in on a periphery and comparing this periphery with its centres might also help us become more sensible towards those practices and processes at the centres that help make and tentatively stabilize vital phenomena as bio-objects. In particular, I argue that the activities and (in)activities of Italian policy makers might facilitate a reading of the 'new modes of governance' (Gottweis et al. 2008, Rose 2007) emerging within and outside of Europe, yet notably—as I will claim—*not* in Italy, neither as universally occurring phenomena of state transformation nor as after-the-fact attempts to tame the consequences of knotty bio-objects, but instead as constitutive and—indeed—generative of these very objects. Yet before venturing into reading between and beyond its lines, let me first tell the story of the public life of Italian IVF embryos by exploring some suggestive moments.[5]

Stage 1: IVF embryos as 'citizen subjects' on parliamentary floors

IVF embryos entered Italian laboratories in the beginning of the 1980s. In 1983 and 1984, news of Italy's first 'test-tube children' was reported (Flamigni and Mori 2005: 14, Neresini 2000: 367, Pace 1984). The language in which journalists spread this news was ambivalent: On the one hand, journalists proudly reported the quick appropriation of this foreign technology by Italian bio-medical professionals, yet the expressions of pride were interspersed with concerns about where these new technologies might lead in the future (Prattico 1984, Repubblica 1985). Over the late 1980s and early 1990s, journalists' concerns shifted from imagined futures to materialized presents. Then, a series of cases of women who had become mothers with fertility experts' helping hands, were controversially discussed in newspapers, on television shows, and in frequent bio-ethical seminars. The female protagonists of most of these stories were women who sought to fulfill their hopes for motherhood capitalizing on the extended reproductive powers of their current family, such as by using eggs donated by female relatives (Giua 1985), or by

---

5   This story is based on an interpretive study (Fischer 2003) in which I empirically followed IVF embryos as they appeared on Italy's political stages (Clarke 2005). Concerning my materials, I did not make a priori decisions on a particular kind of data. I drew on media data from newspapers, television shows and radio broadcasts, on policy documents, on the transcripts of Parliamentary discussions, and on semi-structured interviews that I conducted in Italy with policy makers, bioethicists, scientists and patient activists.

borrowing the wombs of their mothers or sisters (Repubblica 1989); other highly discussed cases involved '*mamme nonne*' (granny moms), i.e., post-menopausal women in their fifties who, so the story went, transgressed the boundaries of responsible and reasonable motherhood by restocking their naturally limited egg supply with the reproductive material of younger women (Grecchi 1994, Repubblica 1992). Pared with the male protagonists of these stories, namely maverick fertility doctors, these female 'transgressors' and 'artificial families' (Repubblica 1990) brought substantial 'family messes' and 'kinship troubles' into an Italian body politic that had already become qualitatively more plural while at the same time quantitatively less prolific. Yet these multiplying realities were nevertheless concealed by an official discourse that condoned the homogeneous shape of 'proper' Italian families (Bernini 2008).

Another reason for these stories was the absence of comprehensive legislation. This 'legal void' (Neresini 2000, Neresini and Bimbi 2000) was the effect of contradictory views on what comprehensive legislation should look like, and on what kind of shared assumptions it could be based upon. The vociferously proclaimed opposition of the Roman Catholic Church toward assisted reproductive technologies made a potential political stage appear even more slippery, in a society in which the Catholic Church does still have a strong say yet is nevertheless no longer universally accepted as a legitimate political actor. Until March 1997, the month in which the birth of Dolly the sheep, the globe's first cloned mammal, was reported, Italian policy makers seemed to prefer to co-exist with these ambiguous or—indeed—'monstrous' (Jasanoff 2005b) realities, rather than to venture into taming these monsters by putting them on the political stage. In the meantime, the disorder spreading from Italy's fertility laboratories was discursively constructed as a 'Wild West' (*Far West*) of reproduction (Neresini and Bimbi 2000). This metaphor, which was used ever more frequently from 1993 on, condensed an ambiguous sense of unease in which a 'left' critique of 'savage markets' merged with more conservative anxieties about female transgressions of natural boundaries (Battistini 1994).

I interpret this 'Wild West' as the socio-political medium in which Italy's IVF embryos could be cultured and grown into 'citizen subjects', i.e., subjects whose well-being the state and its authorities would take care of. Conveying a deep sense of disorder made invasive ordering appear both appropriate and—indeed—urgent. Categorizing IVF embryos as citizen subjects was as much a means of this ordering as it was its outcome.

Once assisted reproductive technologies entered the halls of Parliament and members of Parliament started to work on a joint bill, then on numerous amendments of this bill, those members who vehemently subscribed to what would eventually be enacted as Law 40 did not speak in the name of those infertile or sterile couples who might actually have elected them to office. Rather, they lent their voice to IVF embryos. Politicians framed the embryos as tiny children and as the primary victims of the 'Wild West' of reproduction, in which the wishes of emotionally misguided and therefore uncaring mothers were fulfiled by shrewd fertility doctors. Saving these neglected—if not indeed mistreated—children from infringements by their irresponsible parents and profit-oriented bio-medical professionals became the primary goal of Parliamentarians who argued for Law 40 (Metzler 2007, 2011).

IVF embryos were classified not as objects but as human subjects. Moreover, they were framed as particularly vulnerable subjects because they had no means to speak on their own behalf for their rights. They were in need of Parliamentarians who would, first, lend them their voice to speak up for them, and, second, issue legislation that would safeguard their well-being and proper development.

Assisted reproduction in laboratories was disentangled from natural reproduction in bedrooms and IVF embryos disentangled from fetuses. Even those members of Parliament who argued for restrictive embryo regulations did not dare to re-arrange the rights and freedoms of women over the embryos and fetuses in their mother's wombs, and hence in their 'natural space'; yet they *did* reorder the rights and entitlements of women and prospective parents over those embryos 'out of space' that were dwelling in Italian laboratories. These were disentangled from embryos and fetuses that were dwelling in their mothers' wombs and removed from the discretion of their prospective parents. They ceased to be the private objects of their parents; instead, they became collectivized subjects that were not longer the stuff about which individual citizens were entitled to decide (Metzler 2007, 2011).

Members of Parliament enacted a law that solemnly declared that it would ensure the rights of all involved subjects—including those of the 'conceived'. Yet a juridical reading of the following provisions reveals that Parliament actually ended up privileging the rights of IVF embryos (Dolcini 2009), re-arranging the 'wishes' and 'desires' of prospective parents along a list of rights of IVF embryos. The latter included, first, the right to be created only for the sake of embryo transfer. Law 40's Article 13 states that no more embryos than necessary for establishing a pregnancy and—in any way—no more than three may be produced, that all produced embryos have to be immediately transferred into their mother's womb, and that embryos must not be 'manipulated', 'selected' or frozen for storage. The only 'derogation' or exception to this general ban on embryo cryopreservation is noted as 'serious and documented reason[s] of force majeure that concern the state of health of the woman that could not have been anticipated at the moment of fertilization'. Second, the law enshrined the embryo's right to be born into a neat and clean family structure. It prohibited the insertion of the genes of a third party through gamete donation, made the use of a borrowed womb through surrogate motherhood a punishable offense, and restricted access to IVF technologies to stable heterosexual couples 'in a potentially fertile age' (Marchesi 2007). Third, Law 40 also enshrined IVF embryos' right 'not to be killed', banning embryo research, as well as other research endeavours with human reproductive materialities, such as cloning or the production of chimera.

In sum, the legal birth of Italy's IVF embryos as 'citizen subjects' was an attempt to purify a reproductive field that had become both ambiguous and messy. The IVF embryo as 'citizen subject' was as much the effect of this re-ordering as it was its means. However, as I discuss in the next two sections, Parliament's attempt to make order did not put an end to this story.

Stage 2: The IVF embryo as 'one of us' in ballot boxes[6]

In February 2004, when Law 40 was passed by Parliament after almost seven years of legislative debates, many commentators articulated their relief about the end of Italy's 'Wild West'. However, any assumption that the promulgation of Law 40 marked the end of this chapter of Italian history proved to be rash. Assisted reproductive technologies did not disappear from Italy's political stage; instead, the stages and actors multiplied, with ballot boxes and courtrooms displacing Parliamentary floors, and new actors in the form of the fleshy bodies of adult citizens, scientists and legal experts joining bio-ethicists, clergymen and political representatives. In particular, fully-fledged Italian citizens whose damaged bodies had fallen in the blind spot of the legislators' gaze capitalized on constitutionally granted entrance points in order to make their suffering bodies seen.

The baptism of fire of Law 40 started immediately after its enactment in February 2004, when adult citizens started to criticize it harshly. Among the critics of the law were citizens who deplored that the concern of Italy's legislators for the rights of its unborn citizens infringed upon their rights, hopes and lives. Particularly active in this battle was Luca Coscioni, the president of the Luca Coscioni Association (*Associazione Luca Coscioni*), who was confined to a wheelchair because of amyotrophic lateral sclerosis and who invested his hope for recovery in human embryonic stem cell (hESC) research—a research that under Law 40 Italian scientists could still pursue as long as they enrolled hESC lines from abroad, but for which they were no longer allowed to 'kill' Italian embryos. Coscioni denounced the law as having "trampled on human dignity, disregarded hope" and "sentenced thousands of patients to death" (ADUC 2004). He was joined by other adult citizens, including carriers of genetic diseases who saw themselves deprived of their hopes to become parents of healthy children with the help of pre-implantation genetic diagnosis, by scientists who depicted the law as yet another attack on their scientific freedoms, as well as by other individual and collective actors who might not have embodied the effects of this law in a direct way, but who were nevertheless disturbed by what they framed as a 'genuflection' of the Italian Parliament before the Roman Catholic Church. This heterogeneous public solidified its protest, mobilizing a constitutional granted device that enables the electorate to repeal a law by an 'abrogative referendum' once it has been enacted. It collected enough signatures for five petitions for such abrogative referenda in the summer of 2004. In January 2005, the Constitutional Court approved the four partially-abrogative ones.

---

6 For more comprehensive discussions of the referenda, see Metzler (2007, 2011). For an alternative interpretations, see Corbellini (2007).

However, once the referenda campaigns started in January 2005, Italy witnessed what amounted to a rehearsal of the previous debates on the floors of Italian Parliament. The concerns about trampled rights and the damaged bodies of Italy's adult citizens were backstaged and replaced by a debate on the proper classification of IVF embryos. In particular, a campaign organized by the newly founded Science & Life Association (*Associazione Scienza & Vita*) zeroed in on such a reframing of the stakes. Notably, this question was framed not as the stuff of citizen consultations but as a scientific question instead. This 'matter of concern' was very much a 'matter of fact' (Latour 2004) to which science—luckily— provided clear answers. In fact, once an egg cell, the Italian electorate was taught, was fertilized with sperm, that event gave rise to a new and unique human genome, which contained all the information necessary for its future development into a human being (SATduemila and Scienza & Vita 2005a). This embryo, so the argument went, was not a *potential* human being or person; it was a human being *with potential* or, in short, 'one of us' (*uno di noi*). And 'one of us' was not the stuff of politics, neither of the micro-politics of decisions by prospective parents and their physicians, nor of the macro-politics of ballot boxes.

'Life cannot be put to vote' (*la vita non può essere votata*), was the central slogan of an enlightening campaign in which expert spokespersons of the campaign—and of the embryo—disseminated scientific facts and truths and invited lay Italian citizens to refrain from making judgements about "when life is really life and from when it has a value" (SATduemila and Scienza & Vita 2005b). This invitation was not merely an expression of an earnest conviction; it was also a rather clever trick. Indeed, whilst the collective of opponents of Italy's embryo regime assembled a referendum to make themselves seen and heard, the defenders of this regime mobilized the bias of this instrument. Whilst such a ballot enables laws to be partially or entirely cancelled, the Italian Constitution also demands a minimum of 50% +1 of the Italian electorate to cast its vote for such a ballot to be valid. Hence, the defenders of Italy's embryos preferred not to make themselves dependent on the scientific literacy of the Italian electorate; rather, they put their faith in the documented inert portion of the Italian electorate that always ignores the ballot boxes in such consultations (Uleri 2002). And, in the end, their strategy proved successful. On the two voting days in June 2005, almost three-quarters of the Italian electorate failed to show up to vote. The referendum was ruled invalid, and each Italian IVF embryo was now 'one of us'.

Stage 3: IVF embryos as fetuses' 'little brothers' in courtrooms

Once attempts to undo Law 40 through the abrogative referendum had failed, many commentators expressed their delight about the unexpectedly clear victory. Then, Law 40 seemed to be set in stone, with many framing the inertia of the electorate as a sign not of its estrangement but as its active approval. However, from the vantage point of the present, such reactions proved to be—yet again—

premature. Indeed, after the failure of the referendum, attempts to undo Law 40 and its provisions shifted from ballot boxes to courtrooms. There, couples recurred against the provisions of this 'ordinary' law that, so was the charge, contradicted rights that the higher ranking Italian Constitution had instead granted to them.

The series of legal actions had already begun before the referendum campaigns had even started. A first appeal to courtrooms was made in March 2004, only one month after enactment of the law. A dozen of other cases would follow (Costantini 2010). Most of these cases followed similar patterns: All legal actions involved infertile couples that were carriers of severe genetic conditions. And a stable cluster of advocates pleaded their cases. The infertile couples and their lawyers sued fertility experts and centres that had refused to perform pre-implantation genetic diagnosis, arguing that Law 40 did not allow them to fulfil the couples' demands.

Here, unpacking some technical details might be helpful. Pre-implantation genetic diagnosis (PGD) involves the production of a high number of IVF embryos and removals of one or two cells of these early embryos. Subsequently, genetic tests are performed on these cells to determine whether they demonstrate suspected genetic mutations. The results inform decisions on which embryos are subsequently transferred, and which are instead discarded or frozen (Handyside and Delhanty 1997, Handyside et al. 1990). Law 40 did not mention PGD directly; however, the limitation of a maximum number of three embryos to be produced in one IVF cycle, as well as to the provision that once produced, IVF embryos must be transferred immediately, made PGD technically impossible. Moreover, after the enactment of Law 40, Minister of Health Girolamo Sirchia also issued guidelines that restricted analyses on the state of health of embryos to 'observational ones' only, excluding genetic analyses (Ministro della Salute 2004).

In the various appeals, the couples' lawyers argued that this implicit ban of PGD infringed upon the constitutionally granted 'right to health' of their female plaintiffs. Moreover, they highlighted the contradiction that whilst their plaintiffs were not allowed to choose not to transfer an unhealthy IVF embryo, they were still allowed to interrupt the developmental trajectory of this embryo once this has ceased to be an IVF embryo and was developing in a pregnant woman's body instead. This was possible because back in 1975 the Constitutional Court had declared unconstitutional those (fascist) parts of the Italian penal code that banned abortion as a 'crime against the integrity and health of the stock' (Hanafin 2009: 288-329), reframing abortions as a conflict not between women and the population, but between women's rights and an unborn's right to life, and famously arguing that there was "no equivalence between the right not only to life but also to health owned by who is already person, as the mother, and the protection of the embryo who has yet to become a person" (Corte Costituzionale 1975, Galeotti 2003: 116, Hanafin 2009: 231). Three years later, in 1978, Italian Parliament enacted the 'norms for the social protection of motherhood and on the voluntary interruption of pregnanc[ies]', i.e., the Italian abortion law, which exempts from punishment abortions in the first 12 weeks of pregnancy, as well as afterwards in cases of diagnosed or suspected fetal malformations—for the sake of protecting women's mental and physical health (Repubblica Italiana 1978). Paradoxically, bringing the abortion debate and its logics back into the courtroom helped re-assemble IVF embryos and made their rights slightly less supreme.

A first success was obtained in January 2008 when the Regional Administrative Tribunal (*Tribunale Amminstrativo Regionale*, TAR) of Latium cancelled the sections of the ministerial guidelines that restricted analyses on IVF embryos to observational ones, declaring those guidelines to be 'ultra vires' (Corriere della Sera 2008, Ferrando 2009). Hence, genetic analyses were now legally possible, yet still technically pointless. There was still the law's limit of a maximum number of three embryos to be produced in one IVF cycle, and its provision that demanded that all produced embryos had to be transferred. After the ruling of the TAR, legal actions focused on these provisions (Costantini 2010).

Various courts issued different ordinances and sentences; yet, most of them raised the issue of the constitutionality of Article 14 of Law 40, entrusting the Constitutional Court to speak truth on this issue (Costantini 2010, Dolcini 2010: 4, Turillazzi and Fineschi 2008: 2). And in April 2009, the Constitutional Court declared that parts of this article contradicted constitutional rights and principles. Following a consistent part of the argumentations of the remitting judges (Dolcini 2010: 7), the Court declared, first, that Law 40's limitation on the number of embryos to be produced to a maximum of three was at odds with several constitutional principles. Indeed, this provision did not allow physicians to produce a number of IVF embryos appropriate for individual cases. Instead, it forced physicians to treat what might be wildly different situations with the same predetermined practice. Three embryos, so the Court's reasoning went, might not be enough for older women, yet too high a number for younger ones. Hence, this 'unreasonable' provision ended up discriminating against women of older age, who were forced to repeated hormone stimulations, infringing upon their rights to health, and—in cases of multiple pregnancies—also risked endangering the health of the 'fetus' (Corte Costituzionale 2009). Following the sentence of the Constitutional Court, physicians continued not to be allowed to produce more embryos than necessary for establishing a pregnancy, yet how many embryos were 'necessary' was now up to them to decide (Dolcini 2009). Second, the Constitutional Court also reconsidered the provision that stated that all IVF embryos have to be transferred, cancelling the provision that stated that the only 'derogation' of the general ban on embryo freezing had to be related to women's health and of 'force majeure'. Following the ruling of the Constitutional Court, exclusions from the general ban on freezing embryos for storage did no longer have to 'unforeseeable' in kind. In deciding the number of IVF embryos to produce and which of them to transfer, physicians now had to take into consideration women's right to health. In cases in which considerations of women's health did not allow an embryo transfer, IVF embryos now *had* to be frozen (Dolcini 2010).

Hence, whilst in Italian Parliamentary debates, embryos dwelling in laboratories had been disentangled from fetuses developing in their mother's wombs, in Italian courtrooms, IVF embryos were re-entangled with embodied fetuses, i.e., those entities that courtrooms were by then trained to deal with, and therewith also re-entangled with their mothers. Parliament had treated IVF embryos and fetuses as different entities, the latter being regulated by an abortion act that entitles women to make their own medically surveilled choices on whether they want to become mothers or not, while the former were removed from the discretion of their parents and treating physicians. The Constitutional Court breached the boundary that had only just been erected, re-entangling IVF embryos with embodied fetuses and the legal categories and relationships with which the latter are governed. IVF embryos ceased to be 'out of space', with petri dishes forming a sort of extended wombs over which Italian women enjoyed a medically surveilled sovereignty. Hence, with the sentence of the Constitutional Court, IVF embryos ceased to be different in kind. They became embodied embryos' and fetuses' little brothers and sisters, instead. The sentence did not revoke the list of rights that Parliament had solemnly established; yet, it made them slightly less supreme, balancing the embryos' rights to development with their mothers' right to health.

In summary, when the Italian Parliament had enacted a law in February 2004 that categorized IVF embryos as collective 'citizens subjects' on which individuals were no longer entitled to decide, they re-ordered a reproductive field that had become disturbingly messy. They endowed these new citizen subjects with a list of supreme rights, and re-arranged the rights and freedoms of their parents and bio-medical professionals, accordingly. And yet, once the law had been enacted, adult citizens mobilized different constitutionally granted devices to re-claim rights that Law 40 had stripped from them. Whilst a first attempt to appeal to the Italian electorate had failed, a second attempt to appeal to judges in courtrooms was more successful, when the Constitutional Court declared unconstitutional certain parts of this law. With this ruling, IVF embryos were not transformed into objects. They continued to be highly protected entities. However, bringing the abortion debate back in through legal courts, and re-entangling IVF embryos with embodied fetuses, helped re-assemble them under previous legal realities and made them a little less sacred.

Discussion

What does this story of the making and the unmaking of Italian embryos tell us (or at least hint us at) about the relationship between bio-objects and states, and between knowing and doing lives in labs and governing lives on political scales? In order to answer this question, it is helpful to consider not only what Italian state authorities did, but also what they refrained from doing, especially if their activities and inactivities are compared to the activities of other states.

From this perspective, Italian authorities did not rush into regulating assisted reproductive technologies and the realities that these help generate. Indeed, while many European countries were busily drafting regulations on assisted reproductive technologies, Italian state authorities refrained from drafting such legislation for almost two decades. This contrast is perhaps most striking in comparing Italy to the United Kingdom. In the latter, various state authorities puzzled over how to deal with these technologies and the new realities that these helped generate from the early 1980s onward. These well-documented deliberations culminated in the 1990 enactment of the Human Fertilisation and Embryology Act, as well as the establishment of the Human Fertilisation and Embryology Authority (HFEA), which was entrusted with the implementation of that act (Gottweis, Salter, and Waldby 2009, Mulkay 1997). Moreover, UK regulatory authorities were also careful in drawing boundaries around the kind of entities that could be used for research purposes—and as such 'objectified'—and entities that were not allowed to be used for such purposes. They carefully disentangled early 'pre-embryos' that were amenable to be used for well-defined and later amended research purposes from the more subject-like 'embryo proper' (Mulkay 1994, Spallone 1999). The Italian Parliament, on the other hand, refrained from defining 'conceiveds' and 'IVF embryos' in Law 40 and hence from drawing any boundaries around early forms of human life. It admitted all fertilized oocytes into the moral community of citizen subjects, re-arranging the rights and freedoms of the other members of this community accordingly.[7] This suggests that (state) laws and regulations matter as they enshrine boundaries that help stabilize vital assemblages as researchable or even tradeable bio-objects. They do not come after the bio-object; rather, they are performative in their effect, as they enable epistemological or economic objectifications of vital phenomena. They do not come once vital phenomena are bio-objectified; rather, they make such objectifications possible.

And, yet, the story of the Italian IVF embryo might bear yet another subtle message. The consistent Italian reluctance to regulate assisted reproductive technologies and their products in the past, just as the current reluctance to issue new guidelines in the present, suggests that the absence of legislation has sometimes a similar effect as the presence of highly restrictive ones. Despite the widespread mourning about Italy's 'Wild West', there was no evidence of a burgeoning embryo research industry in Italy before the enactment of Law 40; nor does such an industry seem to exist today, where 'surplus embryos' dwell in a regulatory limbo. Hence, an absence of legislation and regulation might induce an uncertainty that triggers similarly hampering effects for the emergence of bio-objects as the issuing of explicit prohibitions; or in more positive terms: laws and regulations that set clear boundaries around the kind of emergent biologies that can be used for research purposes or that might even be 'tradeable parts', seem to be part of those extended cultures that these emerging life forms need to prosper

---

7   This lack of definition led to juridical and bio-ethical debates, on whether so-called ootide—i.e., an egg cell that has already been 'penetrated' by a sperm cell but where the two cells have not yet fused into a new genome—amounts to an 'embryo', and to a practice in which fertility experts refrained from freezing such entities, freezing unfertilized egg cells instead.

and grow. Certainly, such laws and regulations might make the micro-practices of bio-scientists burdensomly bureaucratic and overloaded with paperwork; yet, they also make these micro-practices possible in the first place.

And, yet, as the story of the Italian IVF embryo shows, not all states are that heavily involved in bio-objectification processes of IVF embryos (or other vital assemblages). Here, unpacking not what Italian legislation did but *how* this legislation was done might be instructive for briefly reflecting upon which states are more likely to be entangled in bio-objectification processes. Indeed, once assisted reproductive technologies and embryos were put on political stages, Italian policy makers were not particularly adventurous. Rather, the practices that were used to govern these matters belonged to a politics 'as we know it'. Citizens were not absent from this politics, yet they had to mobilize constitutional and juridical channels to be granted a say. They were neither invited, nor did state authorities seek to draw upon this extended knowledge to render their ordering practices more robust. Rather, decision making processes were dominated by a Parliament that made a law, among others, by mobilizing the authority of science, which was granted the sovereign right to speak truth on these matters. Here, the referendum was a particular instructive moment of this story. Then, this tacit assumption was vociferously spoken, when many saw as deeply disturbing the very attempt to grant a lay electorate a say on what embryos are and what they should be. In other words, Italy did not only issue restrictive legislation, it also relied on deeply 'modern' practices to issue this legislation. Again, a brief comparison with the UK might be helpful; there, policy-making was not reinvented, yet British authorities made continuous attempts to 'fine-tune' their 'machineries of governance', for instance by creating new regulatory bodies such as the HFEA and by adding on participatory practices to their decision-making processes (Gottweis et al. 2008, Gottweis, Salter, and Waldby 2009). Please note that by continuously referring to the UK, I neither want to suggest the UK as a kind of 'best practice model' for good governance, nor do I want to define Italy as a deviation from that norm. Rather, I think that the UK features among the centres of the geography of bio-objects—embryonic ones as well as of other kinds[8]—and that British' authorities continuous attempts to lubricate their machineries of governance are more than window-dressing exercises. Making life epistemologically, ontologically and economically more productive in labs and making (life-science) governance more efficient, transparent, and democratic seem to be two processes that cannot be reduced into one process; yet, these mutually stabilize and enable each other. Cautiously put in more general terms, it seems that states, such as the UK, that might not necessarily re-invent, but that make efforts to lubricate their machineries of governance, are much more likely to become entangled in bio-objectification processes than states that rely on machineries of governance 'as we know them'. While the

---

8 See Nik Brown's chapter (this volume) for an exploration of the making up of another bio-object—interspecies embryos—in the UK.

latter rely on an understanding of 'life' as something that pre-exists politics and that needs to be purified from politics (Latour 1993)—and hence a sort of Life with a capital L—, the former have the ability to 'secularize' life and to 'open it up' to new ways of knowing and ordering. Such states have an understanding of life as something that is neither out there nor sacred, but as something that men and women are allowed to act upon and to collectively decide upon. And this understanding is performative: It informs ordering practices that inform laws and regulations that enact lives as bio-objects. Just as the 'old machineries of governance' were constitutive for our understanding of Life and Nature as something that is out there, and that transcends our human existence (Jasanoff 2004b), so might re-tuned machineries of governance help generate changing understandings and matterings of life and nature—such as life articulated as tradeable bio-objects.

Conclusions

In this chapter, I have sought to use the uneven regulatory path of Italian IVF embryos to start to think through the relationship between life and politics, and between changing forms of life and changing forms of politics. In a nutshell, I have argued that we should integrate states into our analytical agenda if we wish to make sense of bio-objects and understand their current salience and their uneven geographies. In particular, I have suggested that states are important because some of them (may) issue regulations and legislations that help stabilize vital assemblages as researchable and—may be—even tradeable bio-objects. Moreover, I have argued that unpacking the practices with which states issues such legislation and regulation might be instructive for understanding why some states are more heavily involved in the making of bio-objects rather than others. Indeed, changing modes of governance do not seem to be universally occurring phenomena, nor do they seem to be the consequence of the salience of bio-objects; instead, they seem to feature among those generative conditions that enable the emergence and salience of bio-objects, as 'states' or 'modes of governing (life)' and 'bio-objects' or 'modes of knowing (life)' seem to be co-produced (Jasanoff 2004a).

Yet, does this imply that it is necessarily the state that issues such regulations? A tentative answer to this question is both no and yes. Regulations are often done at a 'sub-political' level, in the form of guidelines or standard operating procedures in laboratories or across laboratories (Cambrosio et al. 2006) and they are also done at an 'international' scale. Indeed, such border-crossing regulations seem to be ever more important for those bio-objects that travel beyond national boundaries, circulating in transnational 'technological zones' (Barry 2001). And yet these transformations do not make states meaningless; rather, they help transform and reconfigure states. How this works out in practice is yet another story—that border-crossing bio-objects might help us to tell.

References

ADUC 2004. *Italia. Luca Coscioni: aboliamo la legge che vieta la ricerca, insieme* [Online: Notiziario Cellule Staminali, Anno III (Numero 57), 9 March]. Available at http://staminali.aduc.it/php_newsshow_0_3059.html [accessed 27 July 2007].

Agamben, G. 1998. *Homo Sacer: Sovereign Power and Bare Life.* Stanford: Stanford University Press.

Associazione Luca Coscioni 2005. *Io voto 4 volte sì.* Roma: Associazione Luca Coscioni.

Barry, A. 2001. *Political Machines: Governing a Technological Society.* London and New York: The Athlone Press.

Battistini, F. 1994. Giovanna Melandri: aberrazione causata dalla mancanza di leggi. *Corriere della Sera,* 26 September, 13.

Beck, U. 1993. *Die Erfindung des Politischen: Zu einer Theorie reflexiver Modernisierung.* Frankfurt am Main: Suhrkamp.

Beck, U. 1997. Subpolitics. *Organization & Environment,* 10(1), 52-65.

Bernini, S. 2008. Family politics: Political rhetoric and the transformation of family life in the Italian Second Republic. *Journal of Modern Italian Studies,* 13(3), 305-24.

Cambrosio, A., Keating, P., Schlich, T. and Weisz, G. 2006. Regulatory objectivity and the generation and management of evidence in medicine. *Social Science & Medicine,* 63(1), 189-99.

Clarke, A.E. 2005. *Situational Analysis: Grounded Theory After the Postmodern Turn.* Thousand Oaks, London and New Delhi: Sage Publications.

Comitato Scienza e Vita 2006. *Referendum 2005 sulla Fecondazione Medicalmente Assistita: Essere umani dall'inizio alla fine. Quattro mesi vissuti intensamente per affermare il primato della vita,* edited by B. Rosati. Pomezia: La Fenice Grafica.

Corbellini, G. 2007. Scientists, bioethics and democracy: The Italian case and its meanings. *Journal of Medical Ethics,* 33(6), 349-52.

Corriere della Sera 2008. Tar Lazio: Legittima diagnosi preimpianto. *Corriere della Sera* [Online, 23 January]. Available at: http://www.corriere.it/cronache/08_gennaio_23/tar_lazio_fecondazione_3e19eaf4-c9cb-11dc-97c6-0003ba99c667.shtml [accessed 5 February 2011].

Corte Costituzionale 1975. N. 27, Sentenza 18 Febbraio 1975. *Gazzetta Ufficiale,* 26 February. Available at: http://www.cortecostituzionale.it/actionSchedaPronuncia.do?anno=1975&numero=27 [accessed 10 February 2011].

Corte Costituzionale 2009. Sentenza N. 151. *Gazzetta Ufficiale, 1a Serie Speciale – Corte Costituzionale,* 13 May. Available at: http://www.cortecostituzionale.it/actionSchedaPronuncia.do?anno=2009&numero=151 [accessed 10 February 2011].

Costantini, M.P. 2010. *Scheda cronologica sui provvedimenti giudiziari più importanti in ordine alle legge 40* [Online]. Available at http://static. ilsole24ore.com/G/GuidaDiritto/binary/11698473.13/11698473.pdf [accessed 27 December 2010].

Dolcini, E. 2009. Embrioni nel numero 'strettamente necessario': il bisturi della Corte costituzionale sulla legge n. 40 del 2004. *Rivista italiana di diritto e procedura penale*, 52, 928-66.

Dolcini, E. 2010. La legge sulla fecondazione assistita oggi, tra giurisprudenza di merito e corte costituzionale. (Relazione tenuta al Seminario 'La fecondazione assistita: Prospettive ed incertezze dopo Corte cost. n. 438/2008 e Corte cost.n.151/2009: Tutela dell'embrione e del feto nel diritto italiano ed il ruolo della giurisprudenza di legittimità civile e penale', CSM, Roma, 17 settembre 2009.) [Online: *Persona e Danno*, 6 January]. Available at http:// www.personaedanno.it/cms/data/articoli/018044.aspx [accessed 29 Decembre 2010].

Douglas, M. 1994 [1966]. *Purity and Danger: An Analysis of Concepts of Pollution and Taboo*. London: Routledge.

Ferrando, G. 2009. Fecondazione in vitro e diagnosi preimpianto dopo la decisione della Corte costituzionale (Relazione tenuta al Seminario 'La fecondazione assistita: Prospettive ed incertezze dopo Corte cost. n. 438/2008 e Corte cost.n.151/2009: Tutela dell'embrione e del feto nel diritto italiano ed il ruolo della giurisprudenza di legittimità civile e penale', CSM, Roma, 17 September 2009) [Online: *Persona e Danno*]. Available at: http://www.personaedanno.it/ cms/data/articoli/018045.aspx [accessed 4 January 2011].

Filippini, N.M. 2004. Il corpo dominato e la personificazione dell'embrione: Una prospettiva storica, in *Un'appropriazione indebita: L'uso del corpo della donna nella nuova legge sulla procreazione medicalmente assistita*. Milano: Baldini Castoldi Dalai, 97-112.

Fischer, F. 2003. *Reframing Public Policy: Discursive Politics and Deliberative Practices*. Oxford: Oxford University Press.

Flamigni, C. and Mori, M. 2005. *La legge sulla procreazione medicalmente assistita: Paradigmi e confronti*. Milano: il Saggiatore.

Galeotti, G. 2003. *Storia dell'aborto*. Bologna: Il Mulino.

Giua, C. 1985. È nata Maria Cristina. È la prima in Europa 'figlia di sua zia'. *Repubblica*, 17 February, 11.

Gottweis, H., Braun, K., Haila, Y., Hajer, M., Loeber, A., Metzler, I.,Reynolds, L., Schultz, S. and Szersynski, B. 2008. Participation and the New Governance of Life. *BioSocieties*, 3(3), 265-86.

Gottweis, H., Salter B. and Waldby, C. 2009. *The Global Politics of Embryonic Stem Cell Science, Health Technology and Society*. London: Palgrave Macmillan.

Grecchi, A. 1994. Madri e padri a 60 anni. *Corriere della Sera*, 6 January, 31.

Hajer, M. 2003. Policy without polity? Policy analysis and the institutional void. *Policy Sciences*, 36(2), 175-95.

Hajer, M. and Wagenaar, H. 2003. Introduction, in *Deliberative Policy Analysis: Understanding Governance in the Network Society*, edited by M. Hajer and H. Wagenaar. Cambridge: Cambridge University Press, 1-30.

Hanafin, P. 2007. *Conceiving Life: Reproductive Politics and the Law in Contemporary Italy*. Aldershot and Burlington: Ashgate.

Hanafin, P. 2009. Refusing disembodiment. Abortion and the paradox of reproductive rights in contemporary Italy. *Feminist Theory*, 10(2), 277-44.

Handyside, A.H. and Delhanty, J.D.A. 1997. Preimplantation genetic diagnosis: Strategies and surprises. *Trends in Genetics*, 13(7), 270-75.

Handyside, A.H., Kontogianni, E.H., Hardy, K. and Winston, R.M. 1990. Pregnancies from biopsied human preimplantation embryos sexed by y-specific DNA amplification. *Nature*, 344 (768-70).

Hoeyer, K. 2010. *(Ex)changing the Body*. Paper presented at the Life-Science-Governance Research Platform, University of Vienna, 16 June.

Hoeyer, K., Nexoe, S., Hartlev, M. and Koch, L. 2009. Embryonic Entitlements: Stem Cell Patenting and the Co-production of Commodities and Personhood. *Body & Society*, 15 (1), 1-24.

Hopwood, N. 1999. 'Giving Body' to Embryos. *Isis*, 90, 462-96.

Hopwood, N. 2007. A history of normal plates, tables and stages in vertebrate embryology. *International Journal of Developmental Biology*, 51, 1-26.

Jasanoff, S. 2004a. Ordering knowledge, ordering society, in *States of Knowledge. The Co-production of Science and Social Order*, edited by S. Jasanoff. London and New York: Routledge, 13-45.

Jasanoff, S. 2004b. Post-sovereign science and global nature. *Harvard University, Environmental Politics/Colloquium Papers*.

Jasanoff, S. 2005a. *Designs on Nature: Science and Democracy in Europe and the United States*. Princeton and Oxford: Princeton University Press.

Jasanoff, S. 2005b. In the democracies of DNA: Ontological uncertainty and political order in three states. *New Genetics and Society*, 34(2), 139-55.

Latour, B. 1993. *We Have Never Been Modern*. Cambridge, MA: Harvard University Press.

Latour, B. 2004. Why has critique run out of stream? From matters of fact to matters of concern. *Critical Inquiry*, 30, 225-48.

Law, J. 2004. Mattering: Or How Might STS Contribute? [Online: *heterogeneities. net*, 28 June]. Available at: http://www.heterogeneities.net/publications/ Law2004Matter-ing.pdf [accessed 3 February 2011].

Law, J. and Mol. A. 2008. Globalisation in Practice: On the Politics of Boiling Pigswill. *Geoforum*, 1(39), 133-43.

Lezaun, J. 2006. Creating a New Object of Government. *Social Studies of Science*, 36(4), 499-531.

Marchesi, M. 2007. *From Adulterous Gametes to Heterologous Nation: Tracing the Boundaries of Reproduction in Italy*. [Online, Reconstruction, 7(1)]. Available at http://reconstruction.eserver.org/071/marchesi.shtml [accessed 19 April 2007].

Metzler, I. 2007. 'Nationalizing Embryos': The Politics of Human Embryonic Stem Cell Research in Italy. *BioSocieties*, 2(4), 413-27.

Metzler, I. 2011. Between Church and State: Embryos, Stem Cells and Citizens in Italian Politics, in *Reframing Rights: Constitutional Implications of Technological Change*, edited by S. Jasanoff. Cambridge, MA and London: MIT Press, 105-24.

Ministro della Salute 2004. Decreto Ministeriale 21 luglio 2004, Linee guida in materia di procreazione medicalmente assistita. *Gazzetta Ufficiale*, 16 August.

Morgan, L.M. 2006. The Rise and Demise of a Collection of Human Fetuses at Mount Holyoke College. *Perspectives in Biology and Medicine*, 49(3), 435-51.

Morgan, L.M. 2009. *Icons of Life: A Cultural History of Human Embryos*. Berkeley, Los Angeles and London: University of California Press.

Moser, I. 2008. Making Alzheimer's disease matter: Enacting, interfering and doing politics of nature. *Geoforum*, 39(1), 98-110.

Mulkay, M. 1994. The Triumph of the Pre-Embryo: Interpretations of the Human Embryo in Parliamentary Debate over Embryo Research. *Social Studies of Science*, 24 (4), 611-39.

Mulkay, M. 1997. *The Embryo Research Debates*. Oxford: Oxford University Press.

Neresini, F. 2000. And man descended from the sheep: The public debate on cloning in the Italian press. *Public Understanding of Science*, 9, 359-82.

Neresini, F. and Bimbi, F. 2000. The Lack and the 'Need' for Regulation for Assisted Fertilization: The Italian Case, in *Bodies of Technology: Women's Involvement with Reproductive Medicine*, edited by A.R. Saetnan, M. Stefania and M. Kirejczy. Columbus, OH: Ohio State University Press, 207-37.

Pace, G.M. 1984. Nata una bimba in provetta con tecnica tutta italiana. *Repubblica*, 24 May, 14.

Petchesky, R.P. 1987. The Power of Visual Culture in the Politics of Reproduction. *Feminist Studies*, 13 (2), 263-92.

Prainsack, B. and Naue, U. 2006. Relocating health governance: Personalized medicine in times of 'global genes'. *Personalized Medicine*, 3(3), 349-55.

Prattico, F. 1984. Rischiano malformazioni i bimbi nati in provetta? *Repubblica*, 23 December, 14.

Rabinow, P. 1999. French DNA. Trouble in Purgatory. Chicago, IL and London: University of Chicago Press.

Repubblica 1985. Bimbi in provetta. Da oggi a Palermo li 'passa' la USL. Già 2000 domande. *Repubblica*, 14 February, 15.

Repubblica. 1989. Salerno: Partorisce il bimbo della sorella. *Repubblica*, 30 November, 20.

Repubblica. 1990. La 'famiglia artificiale' in cerca di nuove leggi. *Repubblica*, 16 September, 19.

Repubblica. 1992. 'Sarò mamma a 62 anni grazie a un ovulo donato'. *Repubblica*, 23 April, 20.

Repubblica Italiana 1978. Legge 22 Maggio 1978, n. 194, Norme per la tutela sociale della maternità e sull'interruzione volontaria della gravidanza. *Gazzetta Ufficiale*, 22 May.

Repubblica Italiana. 2004. Legge 19 febbraio 2004, n. 40, Norme in materia di procreazione assistita. *Gazzetta Ufficiale*, 24 February.

Rose, N. 2007. *The Politics of Life Itself: Biomedicine, Power, and Subjectivity in the Twenty-First Century*. Princeton and Oxford: Princeton University Press.

SATduemila & Scienza & Vita. 2005a. *Dalla diagnosi prenatale alla diagnosi preimpianto*. Roma: Rete Blue S.p.A. (videotape).

SATduemila & Scienza & Vita. 2005b. *Sperimentazione sull'embrione: Diritti del concepito e salute della donna*. Roma: Rete Blue S.p.A. (videotape).

Sørensen, E. and Torfing, J. (eds) 2007. *Theories of Democratic Network Governance*. New York: Palgrave Macmillan.

Spallone, P. 1999. How the pre-embryo got its spots. Paper presented at the 'Genetics and Genealogy' Conference, Potsdam, 4-6 July 1999.

Turillazzi, E. and Fineschi, V. 2008. Preimplantation genetic diagnosis: a step by step guide to recent Italian ethical and legislative troubles. *Journal of Medical Ethics*, 34(10), e21.

Uleri, P.V. 2002. On referendum voting in Italy: YES, NO or non-vote? How Italian parties learned to control referendums. *European Journal of Political Research*, 41 (6), 863-83.

Vries, G. de 2007. What is Political in Sub-politics? *Social Studies of Science*, 37(5), 781-809.

Chapter 11

Growing a Cell *in Silico*: On How the Creation of a Bio-object Transforms the Organisation of Science

Niki Vermeulen

Visions of the cell appeared in the 17th century when Robert Hooke observed the basic structure of cork through a microscope and noticed spaces that were similar to the small rooms monks used to live in, so-called 'cella' (Sloterdijk 2005, Hooke 1665/1987). Since then, the cell has become an important unit of study leading to the establishment of cytology – currently known as cell biology – and an increasing understanding of the cell (Maienschein 1991). More than 450 years after its discovery, scientists are now rebuilding a cell *in silico*: a replica of a living cell in a computer:

> A Silicon Cell is a precise replica of (part of) a living cell. It is based on experimentally determined rate laws and parameter values, i.e. only on data, not on fitted values or assumptions. It merely calculates the system biology implications of the molecular properties that are already known.[1]

The Silicon Cell does not resemble Hooke's monastery room at all, nor does this cell look like common textbook images of cells. As it consists of an enormous amount of letters and moving lineages forming graphs on a screen representing specific cellular processes, the silicon cell will redistribute, transform and reconfigure the notion of life.

The Silicon Cell can be seen as a bio-object. Its creation involves recent developments in the life sciences and especially the reconfiguring of life as information. On the one hand, investigations into the molecular basis of life have resulted in a wealth of messy information on the cell, its constituents and processes. On the other hand, the informational turn in biology that came together with the development of bioinformatics, enables the ordering, integration and computation of information on the cell. The silicon cell is a model of a cell in a computer, envisioned to mimic all the processes of an *in vivo* cell and to replace living cells in scientific experiments, making research faster, less complicated and more reliable.

---

1   Retrieved: 30 April 2007 from http://www.siliconcell.net.

The Silicon Cell is part of the digitalisation of life that leads to bio-objectification. The construction of a model of a cell relates to initiatives that aim to build even more complex entities of life *in silico*, like a virtual plant, a virtual heart, a complete virtual human and even the replication of whole ecosystems on a screen. These virtual models of life are specific forms of bio-objects, as their attribution to live is only derived from the living object they represent. The development of these bio-objects interacts with a shift from 'wet research' in the laboratory to 'dry research' using information technologies (Penders et al. 2008). Following this well-known vocabulary the bio-object at hand is a 'dry bio-object'.

Moreover, the modelling of organisms is part of the movement towards 'systems biology' in which life is defined as a system of relations. While genomics and other -omics research primarily focuses on the analysis of separate parts of the cell, systems biology aims to make sense out of the huge amount of data through the integration and contextualisation of information in models. This can be seen as a second shift, from defining life as information to defining life as a system of relations. So the Silicon Cell as bio-object is a configuration of relations that represents and calculates life.

This chapter will especially focus on the way in which the creation this bio-object transforms the organisation of science. In other words, the process of bio-objectification also entails a process of reorganising science. While the project starts out as an effort of a small group of scientists, they need to orchestrate the European research community towards a standardised version of a Silicon Cell. They present the construction of a Silicon Cell as so-called 'big science', as the analysis and integration of all the different parts of a complex cell requires a huge effort that will be even bigger than the Human Genome Project.[2] By analysing the building of the virtual model, I will argue that the creation of this bio-object requires the construction of large-scale international scientific collaboration and the transformation of government arrangements in science.

In my analyses I distinguish three different phases in the process of developing the Silicon Cell. The first phase of bio-objectification 'Visions of a bio-object' will look into the origins of the Silicon Cell Initiative and the construction of the Silicon Cell as a bio-object. Subsequently, 'Staging a bio-object as big science' examines how the Silicon Cell Initiative is put forward as large-scale science. However, as this did not result in actual funding for the initiative, a change of strategy followed consisting of a turn towards systems biology. This forms the third phase. Finally, I will review how the construction of the Silicon Cell bio-object requires the construction of large-scale international scientific collaboration and the transformation of government arrangements in science.

---

2   The term 'big science' focuses on increasing scales in science, measured in terms of money, scientists, publications and size of instruments used. The term emerged in the 1960s in the context of physics research and has subsequently been used to study various large-scale scientific efforts (Vermeulen, 2009).

My analysis of the Silicon Cell Initiative and the systems biology efforts discussed in this chapter is based on information about SiC and systems biology, consisting of websites, policy documents, meeting minutes, scientific articles and books. In addition, I have performed in depth interviews with scientists and policy makers and I have observed meetings dedicated to systems biology in the period from 2005 until 2009.

Visions of a bio-object

When asked about the origin of the Silicon Cell Initiative, Professor Hans Westerhoff, head of the department of molecular cell physiology at the BioCenter Amsterdam and the key person behind the initiative, points directly at PhD research of two of his former students. But then he hesitates and starts reflecting by asking: "How does something like this start?" (interview Westerhoff 2005).[3] It turns out to be quite difficult to trace back the emergence of the initiative. However, gradually the history of the Silicon Cell is reconstructed from bits and pieces of Westerhoff's memory, complemented by the view of other participants and documentation about the initiative. It turns out that not only PhD students, but also yeast and sleeping sickness play an important role in the origin of the bio-object, as well as the overall goal of finding a cure for cancer. Moreover, I will show how a scientific dispute, a crucial experiment and going public are key elements in the project of bio-objectification.

Research as the basis for bio-objectification

In the beginning of the 1990s, doctoral research on three different organisms – yeast, *T.brucei* and *E.coli* – constituted the beginning of the Silicon Cell. First of all, Bas Teusink investigated sugar decomposition in a yeast cell and found oscillation as mechanism behind this process: glycolytic oscillations. However, another scientist from another group came with a different mechanism which could also explain the phenomenon. This resulted in a heated academic debate, which made Westerhoff think of ways to solve the dispute: "This moment stirred me up as I thought: this is not possible. One person fully defends one mechanism by shouting and railing against other people who have a different model and actually both parties do not have a leg to stand on" (interview Westerhoff 2005). Westerhoff figured that the only way to solve the dispute was to make a precise model of the process in the cell. This showed that a combination of the two mechanisms was in place. So, it

---

3   Prof. Dr Hans V. Westerhoff is leader of the Department of Molecular Cell Physiology at the Faculty of Earth and Life Sciences of the Free University in Amsterdam. In addition, he is a group leader for systems biology at the Manchester Centre for Integrative Systems Biology (MCISB) in the Manchester Interdisciplinary BioCentre (MIB), which is part of Manchester University.

was an attempt to solve a scientific dispute in the context of PhD research that led
to the building of the first Silicon Cell.

Subsequently another doctoral student and another organism entered the stage,
presenting another route towards the Silicon Cell. Barbara Bakker investigated
Trypanosoma brucei in order to find an effective way to tackle the parasite that
causes sleeping sickness. Bakker's research tried to find a way to kill the parasite
by inhibiting a step in its metabolism: "In other words, you have to find a place
on the metabolic highway where you can put a stone so the cars cannot drive
anymore" (interview Westerhoff 2005). The challenge is to find a crucial step or
enzym that can be inhibited but does not appear in the metabolism of a human cell.
In this way, the drug will only kill the parasite and will leave the human cells intact.
The building of a model made it possible to calculate which step this needed to be:
"A fantastic result" (idem). Finally, the third project of Johann Rohwer, on sugar
decomposition in the bacterium *E-coli*, also successfully employed the modelling
technique. If these doctoral research projects provided the foundation for the
Silicon Cell Initiative, it was a crucial experiment that solidified the potential of
the Silicon Cell as a new bio-object, as the next section will explain.

Articulating the bio-object

While the idea of the Silicon Cell implicitly took shape within the context of the
different lines of PhD research at the beginning of the 1990s, it became explicit
in 1997. At that time the bio-object definitely proved to work: "At that moment
we were able to use the calculations to find a drug target for *T.brucei* and we also
accidentally discovered a biological phenomenon, the functionality of something"
(interview Westerhoff 2005). More precisely, the Silicon Cell played a crucial role
in the discovery of the function of the organel in *T.brucei* and yeast cells. These
cells are unique because they have an organel around the enzymes that decompose
sugar. By using the computer model, the scientists were able to see what happens
when removing this organel: the cells explode. According to Westerhoff this
experiment could not have been done in a real cell, as the removing of one part
would have deranged the complete cell. Therefore the experiment proved the use
of the computer model, which was a crucial step in the development of the Silicon
Cell Initiative: "At that moment you have a model that works, you lay your hands
on something with which you can go public" (idem).

Going public meant communication of the scientific observations and results
to fellow scientists, which led to the articulation of the idea of the Silicon Cell.
"You are enthusiastic about the results and you start telling stories about it"
(interview Westerhoff 2005). The stories were followed by lectures, invitations to
conferences and writing results down: "And this is where you start thinking about
the wider significance because that is important when you tell your story: why is
it interesting for other people?" (idem). The process of presenting the research
results made Westerhoff think about the Silicon Cell as a principle or practice that
should be applied more widely. As a result he started to mention the Silicon Cell

in articles and he constructed the Silicon Cell website together with Jacky Snoep, who had worked in Amsterdam before becoming a professor at Stellenbosch University in South-Africa.

Contextualising the bio-object

In the presentation of the bio-object to others, its context and use of become important. Work on the Silicon Cell and the interest in the functioning of cells stems from an interest in improving human health. "We were actually interested in tumour cell biology and the development of drugs, but that turned out to be too difficult because we did not know the system well enough" (interview Westerhoff 2005). 'The system' refers to the way in which the elements of a tumour cell interact. Every cell consists of various elements that independently of each other do nothing. Only in interaction these different parts make up the processes in a living cell, like the cyclical process of cell division. So when wanting to control processes in a tumour cell in order to fight the cancer, you have to figure out which enzymes can influence the processes in the cell most effectively. This can be done by modelling: the Silicon Cell.

As the modelling of cancer cells is very complex, another less complicated cell had to be chosen to begin with: "So then we started thinking about which system we could use to start constructing these precise models and that turned out to be yeast" (interview Westerhoff 2005). The choice for baker's yeast, or *Saccharomyces cerevisiae*, was both scientifically and economically interesting. Yeast is relatively simple and well-known as it functions as a model organism in biology and it is fundamental to various important industrial processes like baking and brewing. As a result, a small group of scientists around Westerhoff started to investigate and model the mechanisms in the yeast cell. "However, the final goal is still the tumour cell and there the problem is of course that this cell is almost similar to the cells of the host" (idem). This makes the case of cancer much more complex, but the ultimate aim is the building of precise models of a cancer cell to find ways to effectively inhibit vital processes in these cells.

Networking the bio-object

The idea of the bio-object got shape within a growing network of researchers. The Silicon Cell emerged within the traditional academic research group of Westerhoff, which consisted of about eight people at the end of the millennium. This network soon expanded as interaction with other scientists took a central place in the development of the idea of the Silicon Cell. Although the idea of the Silicon Cell bio-object did emerge in the interaction between scientists, those relations did not have the form of a structural collaboration yet. Only after its articulation the idea of the Silicon Cell became a starting point for further organisational activity.

Staging a bio-object as big science

In the second phase of the Silicon Cell Initiative, the bio-object becomes the central theme of several organisational activities aiming to build a large-scale research project, and the initiative evolves towards the aim of building a formal collaboration. With the Silicon Cell website as a starting point, I will discuss the Amsterdam Silicon Cell consortium and proposals for research collaborations on a Dutch and European level in which the Silicon Cell Initiative grows and is put forward as big science (De Solla Price 1963, Weinberg 1967). The analyses of this project in development illustrates that the elaboration of ideas and organisational matters continues to be interwoven as they were in the first stage.

Building a website

Together with Snoep from the group at Stellenbosch University, Westerhoff constructed a website entitled 'SiC!: The Silicon Cells' which can be found at the special domain siliconcell.net. The website forms the virtual heart of the initiative and exists of general information about the Silicon Cell, a separate website with the actual models and an overview of organisational activities.

The main page presents the Silicon Cell to a wider public by giving a precise definition of the Silicon Cell and positioning it against other cell modelling initiatives, or other bio-objects. Making a model of a cell is the central theme of different projects scattered around the world: like the e-cell from the Institute for Advanced Biosciences at Keio University in Japan and the Virtual Cell project in the United States. In contrast to these projects, the Silicon Cell is defined as a *replica* and not as a *simulation*:

> A simulation can be a replica but in practice it often is not, as people just take a model and try to make it simulate the real situation by adjusting some parameters. That is absolutely forbidden in the case of our replica: we are not doing that. We are only allowed to change a parameter if we measure that in the experimental situation because we want to explain the behaviour of the system from the molecular behaviour. (interview Westerhoff 2005)

In addition to demarcating the Silicon Cell from similar initiatives, the website present results of research by linking to a separate website: the 'Silicon Cell ready to use: the website with Silicon Cells that can be run over the web'. The website has two objectives: putting the available models on the web to make them accessible for others and tempting scientists from all over the world to add new models to expand the Silicon Cell Initiative. As a result, the database encloses the result of research on processes in various cells and is therefore the materialisation of the Silicon Cell initiative.

Finally, the Silicon Cell website outlines the plans for the building of a Silicon Cell initiative. The Silicon Cell wants to be an open initiative "Everybody can do a small piece of the puzzle, which will together make the bigger picture. As we can never do it alone and I think nobody can" (interview Westerhoff 2005). Therefore the website also explicitly wants to introduce scientists to the Silicon Cell and its uses and offers instructions for using it in the short course 'Playing with silicon cells' (Westerhoff n.d.). Moreover, information on the organisation of the initiative can be found: the Amsterdam Silicon Cell programme that is seeking to expand internationally.

The Amsterdam Silicon Cell programme

The Amsterdam Silicon Cell programme outlines the Dutch research programme, which is a collaboration between the BioCentrum Amsterdam, the Center for Mathematics and Informatics (CWI) and the Institute for Informatics of the University of Amsterdam. The Silicon Cell programme consists of a calculation and an experimental part. The calculation part has as a long-term goal "the computation of Life at the cellular level on the basis of the complete genomic, transcriptomic, proteomic, metabolomic and cell-physiomic information that will become available in the forthcoming years". In other words, this part wants to build the actual Silicon Cell. However, as the completion of this ambition is expected to take more then a decade the work will first concentrate on three major challenges: dealing with the systematic handling of the relevant data and results, networks, space and time.

In addition, the experimental part of the research aims to add to the understanding of processes in the cell, more specifically the function of genes from which the function is not clear yet: "The availability of complete genomes has identified many genes of which the function is unknown, uncertain, or unproven. In many cases this is because the phenotype of these genes is absent, weak, or indirect; we call these the (silent and) whispering genes. Much of ultimate function resides at the flux and metabolite concentration ('Metabolome') level". The experimental part inspects the functioning of large numbers of these genes systematically at the level of metabolism within the context of various small running projects that concentrate on the understanding of single-cell organisms yeast and *E. coli*.

The Amsterdam programme outlines the basic research plans of the Silicon Cell programme. Although the actual research in the Amsterdam programme deals with small organisms and researchers are working on different, relatively small grants, further research plans want to expand research efforts. The Silicon Cell initiative is staged explicitly as research with a big science ambition in programme documents, research proposals and in the stories of scientists involved. As a result, a gap between research practice and research plan emerges: while the actual science is still small-scale, the ambitions of research plans become big.

Towards an international bio-object

The aspiration of bigness was first outlined in the programme entitled '*SiC!* A Dutch initiative for an international *Silicon Cells* program' (Snoep et al. 2003).

The subtitle 'towards a Dutch international Silicon Cells initiative' summarises the aim of the piece:

> This memorandum aims to boost the Dutch momentum behind what should become an international program in an area of molecular system biology that is entitled Silicon Cells (*SiC*). This initiative should result in a major proposal in the framework 6 program of the European Union, in the European Science Foundation, and in the Human Frontier and Science Program. The wider ambition is to set up an international program in which the activities in Europe, the United States, Japan and South Africa are harmonized. (Snoep et al. 2003).

By writing this memorandum, the Amsterdam consortium planned for international expansion, aiming to make the bio-object and its construction an international affair.

The scientists legitimate the enlargement of their research with references to the Human Genome Project that is often seen as the first large-scale project in biology. While the Human Genome Project was only concerned with one part of the cell, the Silicon Cell involves the integration of information on all parts of a cell, which asks for a bigger scientific effort. Accordingly, the Silicon Cell Initiative is presented as a multi-disciplinary collaboration that integrates classical approaches in the biomolecular sciences with bioinformatics, genomics, proteomics, and metabolomics. The integration of all the available information on a cell into one model simply cannot be done in one lab or in one country. Consequently, European collaboration and collaboration with industry is required to acquire the critical mass that is able to create a Silicon Cell of a more complex organism within a reasonable timeframe.

Moreover, it is argued that central coordination of research is needed enabling the standardisation necessary to develop a functioning Silicon Cell model. "Currently, lots of people are working on models but this is organised on a national level. In the end, this will bring various models – a model here and a model there – but these models will not fit together and that's a big shame" (interview Westerhoff 2005). For instance, for the combination of different models temperature is a crucial factor:

> When you have a piece of a model that is done for an organism by 25 degrees and another piece that is done at 37 degrees then they will not fit. Although the temperature maybe does not matter for the structure it just will not work when you put the two pieces of model together, which is frustrating. When they both would have been done at 27 degrees, you would not have a problem at all. (idem)

So in order to be able to put the work on models in different countries together in the end, first an agreement has to be made on temperature, as well as on other standards that are necessary to make different parts of the model compatible.

Finally, the internationalisation of the initiative is legitimised economically by addressing the relation with industry. The pharmaceutical and food industry are interested in research into the working of cells in general and especially in the working of certain specific cells, like the liver cell or yeast cell. Research into the yeast cell is interesting for companies as DSM and Unilever as the cell plays an important role in industrial processes like brewing and baking. The liver cell finds use in the medical realm, calculating the effects and conversion of drugs in the liver as well as liver regeneration after hepatitis and alcohol abuse. It is even argued that the medical world will be transformed by the shift from a largely empirical to a calculation-based operation. By outlining the applications of the bio-object, it is argued that the initiative will have an economic impact: "It should be clear that the economic importance of the SiC initiative is substantial both for the near and the more distant future" (Snoep et al. 2003).

In conclusion, two ways of legitimising the bigness of the Silicon Cell initiative have become visible. On the one hand a scientific line of reasoning can be observed, taking recent scientific and technological developments as a basis for the positioning of the Silicon Cell bio-object as reasonable objective when forces are joint and coordinated. On the other hand the need for a international research network is outlined in a policy context. Building on the Human Genome Project that transformed biology into big science, the Silicon Cell initiative is put forward as a logical next step, delivering new scientific insights and important economic benefits.

The bio-object as big science

While in the first phase the articulation of the bio-object within the research context took place, the second phase stages the bio-object as big science. The Silicon Cell is presented as a unique and innovative concept to a wider public, mainly consisting of other scientists, policy makers and industry. Although the research is still small-scale and only first results can be shown, the research plans put the Silicon Cell forward as big science and its relevance is outlined in a science policy context as well as an industrial context. As these plans did not result in funding for an actual European or international research project on the Silicon Cell, Van Driel and Westerhoff started to follow their pursuit on another level: promoting the broader research trend towards systems biology in which the Silicon Cell initiative can be embedded.

The turn towards systems biology

In 2005, Van Driel says: "forget the silicon cell, now it is systems biology" (interview Van Driel 2005).[4] From 2000 onwards the term 'systems biology' has become very popular in biology and can be seen as the new trend after genomics

---

4   Prof. Dr. Roel van Driel is biochemistry professor leading the group Structural and Functional Organisation of the Cell Nucleus at the Swammerdam Institute for Life

(Bock von Wülfingen 2009, Fujimura 2005, Fox-Keller 2005, Webster 2005). In March 2002 *Science* dedicated a special section to systems biology and *Nature Biotechnology* followed with a special issue in 2004.[5] The development towards systems biology is often conceptualised as a turn from a reductionist to a holistic approach in biology (Chong and Ray 2002). While genomics and other -omics research primarily focuses on acquiring information about the different parts, systems biology aims to make sense out of the huge amount of data through integration of information in models.

> Systems biology is about the iterative cycle of integrating data from a database into a model. Subsequently, the model can be used to make a hypothesis, which can then be tested in an experiment. The results of this experiment can then be fed back into the model again, which creates an iterative cycle that makes the model better and better. (interview Rietveld 2006)[6]

Nowadays, systems biology has all the characteristics of a discipline: its own champions, its own journals and a yearly international conference on systems biology is organised. After systems biology champion Leroy Hood established the first Institute for Systems Biology in Seattle, similar institutes are now appearing all over the world. In addition, systems biology is often seen as a way to acquire research funding: "Some people view systems biology as another complot of the molecule-mafia to acquire money" (interview Rietveld 2006). This is underlined by an article in *The Scientist* (Wiley 2006) that invites its readers to follow the hype by describing 'Five simple steps towards a systems biology approach'. Moreover, systems biology is presented as a scientific enterprise that will deliver profits (Mack 2004). It will influence all fields where live organisms or processes can play a role, varying from health, agriculture and food to industrial processes, energy and the environment (Remacle and Benediktsonn, 2006).

As the Silicon Cell and other modelling efforts can be seen as part of this broader development, Westerhoff and Van Driel started to proliferate their work as systems biology. They turned their attention to the promotion of this new approach within academia and towards policy on a Dutch and European level, in order to support their work on the modelling of cells.

Sciences of the Faculty of Science of the University of Amsterdam. In addition, he has been appointed Faculty Professor at the Faculty of Science.

   5   *Science*, 295, 1 March 2002; *Nature Biotechnology*, 22, October 2004.

   6   Dr. Luc Rietveld was at the time working as a policy officer for ZonMW and the Netherlands Genomics Initiative and dedicates part of his time to coordinating systems biology research.

Performing systems biology inside academia

To encourage the new development towards systems biology that emerged at the beginning of the new millenium, Westerhoff and Van Driel engaged in legitimising and organising systems biology within a Dutch and European scientific context. While aware of the contested nature of the concept, the two scientists underlined their belief in systems biology with words and actions. Several articles on systems biology have been written, such as a publication in *Nature* in 2004 entitled 'The evolution of molecular biology into systems biology' that Westerhoff wrote together with Bernard Palsson, professor of Bioingineering at the University of California in San Diego (UCSD). In addition, Westerhoff co-edited a handbook on systems biology. Next to this definition work, various actions underlined the importance of systems biology. Westerhoff and Van Driel started to talk to various prominent Dutch scientists: "We just made an appointment and asked how they thought about current developments" (idem). Also, a new website was built, available at: http://systems-biology.net. Now this website is the umbrella that spans national and European activities in systems biology, including the Silicon Cell website. In addition, various workshops and conferences have been organised on a Dutch and international level as well as educational activities. The various articles, books, courses, workshops and conferences, which come together on the new website, have certainly put systems biology on the scientific agenda.

Performing systems biology outside academia

If scientists had to be convinced of the new approach, systems biology also needed to earn a place for itself on the science policy agenda. Westerhoff and Van Driel contributed to the development of policy for systems biology on a Dutch and European level. They performed a 'Forward Look on Systems Biology', which resulted in the report entitled *Systems Biology: a Grand Challenge for Europe* (ESF 2005), presenting 'A European Action Plan' for the creation of a European Systems Biology programme. The plan consists of seven actions geared to building an infrastructure for systems biology research in Europe: the set-up of a task force to develop a European road map; establishing a consortium of European Reference Laboratories; cooperation between industry, academia and charities; public acceptance; training and education; financing; and the establishment of a European Systems Biology Office (ESBO). Especially the financing of this whole systems biology enterprise is seen as an important challenge:

> The Europe-wide approach proposed here will require a higher level of funding than that provided by vehicles available now. We propose that a new financial model is developed based on cooperation between national, international, industrial, charities and EC-related organisations. It will be a major challenge to

> align this heterogeneous set of parties so as to produce a synergistic cooperative programme. (ESF 2005: 6)

In other words, the European Systems Biology programme is of a different scale than current scientific endeavours in biology.

Although this new order is not a reality yet, the scientists continue working on making it happen. They are part of the ESF Task Force on Systems Biology that is trying to pursue recommendations of the ESF Forward Look, establishing ESBO as well as a Grand Action on Systems Biology (GRASB).

> This programme should consist of a portfolio of coordinated activities aimed at an integral activity entitled 'Networks for Life'. The portfolio of activities should build on the major Systems Biology activities that the taskforce members have put in place already, and constitute the world's largest and most effective single Systems Biology programme. (Taskforce on Systems Biology 2007: 5)

In addition, Van Driel and Westerhoff are involved in the design of a European Research Era for Systems Biology. ERASysBio aims to create a systems biology policy that integrates national systems biology initiatives: "The countries involved in ERASysBio are fully aware of the need to join forces in creating international policy on systems biology, to enable competitiveness in the field" (ERASysBio Partners, 2007: 4). In sum, Van Driel and Westerhoff have been continuously working on the construction of a large-scale collaboration in systems biology in order to realise the modelling of cells or the creation new bio-objects.

Conclusion

When analysing the construction of a bio-object in detail, it becomes apparent that the bio-object and the organisation of science evolve together. The first phase demonstrates the close relation between science and its organisation by showing how the Silicon Cell is gradually shaped, simultaneously as a scientific object and as collaboration. The bio-object and its meaning evolve through the different stages. In the first phase, the Silicon Cell comes into being as a means to solve a dispute and a method to perform research. However, it gradually becomes an articulated and stable object that makes a transition from wet to dry biology possible. Next to being a concrete model of (part of a) cell in a database, the Silicon Cell bio-object becomes a vision. In the second phase, the idea of the Silicon Cell is turned into a subject of large-scale collaboration and it acquires various applications. The bio-object becomes a means to improve human health – for instance curing cancer – and it also turns into an industrial tool. In the third phase, the cell becomes part of the broader development towards systems biology. In sum, the meaning and the embedding of the Silicon Cell bio-object changes continuously.

The different definitions of the bio-object are related to different organisational phases. In the first act, the idea of the Silicon Cell emerges out of research practice in a traditional academic environment; with PhD students performing their research and collaborating within a research group headed by a professor. However, when the research results are presented to various audiences – including science, policy and business – the idea of the Silicon Cell becomes articulated and staged as big science. As subject of large-scale collaboration, the network around the Silicon Cell starts to grow and collaboration expands the boundaries of the research group. First, expansion takes place along existing relationships and scientists from other groups, departments and institutes become involved. Later on, new relationships are established and collaboration with policy makers and industry takes shape. In this context also a transition from the specific Silicon Cell initiative towards the broader systems biology approach takes place. In this way, the organisation of the research changes along with the idea and meaning of the bio-object, moving from small-scale research to a large-scale international network. So the analysis of the Silicon Cell Initiative does not only show the construction of a bio-object, but also the transformation of the way in which science is organised and governed.

The Silicon Cell bio-object contributes to the creation of 'big biology'. While some argue that biology is not big science as the size of technologies do not even come close to the size of instruments such as particle accelerators in high-energy physics, the construction of this and other bio-objects show that biology is a networked form of big science that emerged through the integration of information and communication technologies and biosciences. The large-scale research networks in which bio-objects are often embedded have important implications for scientific practice. For instance the role of scientists changes in the construction of the bio-object. Westerhoff and Van Driel transform from quite normal professors running their own lab into drivers of an international scientific endeavour, fulfilling the roles of network manager, lobbyist, policymaker and negotiator. As a result, the Silicon Cell bio-object digitalises relations of life, expands relations in science, and changes ways of doing science. As such, changing relations of life go hand-in-hand with the transformation of relations in science.

References

Alberghina, L. and Westerhoff, H.W. 2002. *The Yeast Silicon Cell: A Molecular Systems Biology Approach: Information Memorandum for an Expression of Interest for an Integrated Project*. [Online: Siliconcellnet]. Available at: http://www.siliconcell.net/ysic/eoimemo.html. [accessed 29 August 2008].

Bock von Wülfingen, B. 2009. Is There a Turn to Systems Approaches in Life Sciences? *EMBO Reports*, 10, 37-42.

Boogerd, F.C. and Westerhoff, H.V. 2007. *Systems Biology: Philosophical Foundations*. Amsterdam: Elsevier.

Chong, L. and Ray L.B. 2002. Systems Biology – Whole-istic Biology. *Science*, 295(5560), 1661.

Driel, R. van 2005. *SysBioNL: A Dutch Programme for Food and Pharma*. Amsterdam: unpublished report.

Driel, R. van and Westerhoff, H. 2003. Systeembiologie levert al fantastische resultaten op. *Bionieuws*, 2003(18).

ESF 2005. *Systems Biology: A Grand Challenge for Europe*. Strasbourg: European Science Foundation.

Fujimura, J.H. 2005. Postgenomic futures: Translations across the machine-nature border in systems biology. *New Genetics and Society*, 24(2), 195-225.

Goodman, N. 1978. *Ways of Worldmaking*. Indianapolis: Hackett.

Hooke, R. 1987. *Micrographia: Or Some Physiological Descriptions of Minute Bodies Made by Magnifying Glasses, with Observations and Inquiries Thereupon*. Lincolnwood, IL: Science Heritage.

Keller, E. Fox 2005. The century beyond the gene. *Journal of Biosciences*, 30(1), 3-10.

Knorr-Cetina, K.D. 1999. *Epistemic Cultures: How the Sciences make Knowledge*. Cambridge, MA: Harvard University Press.

Mack, G.S. 2004. Can complexity be commercialized? *Nature Biotechnology*, 22, 1223-9.

Maienschein, J. 1991. Cytology in 1924: Expansion and Collaboration, in *The Expansion of American Biology* edited by K.R. Benson and J. Maienschein. New Brunswick: Rutgers University Press, 23-51.

O'Malley, M.A. and Dupré, J. 2005. Fundamental issues in systems biology. *BioEssays*, 27, 1270-6.

Price, D.J. de Solla 1963. *Little Science, Big Science*. New York: Columbia University Press.

Remacle, J. and Benediktsson, I. 2006. *Proposal Workshop on Systems Biology*. Available at: ec.europa.eu/research/biotechnology/ec-us/docs/remacle_20_july_12-42_en.pdf [accessed 8 June 2008].

Sloterdijk, P. 2005. *Inspiration*. Paper to the lecture series: Breath-taking. Air, art, architecture, Jan van Eyck Academie (Maastricht, 31 May).

Snoep, J.L., Driel, R. van and Westerhoff, H.V. 2003. *SiC! A Dutch initiative for an International Silicon Cells Program*. [Online systemsbiologynet]. Available at: http://www.systemsbiology.net/sbnl/-TSB.htm [accessed 30 April 2007].

Taskforce on Systems Biology 2007. *ESF Taskforce on Systems Biology: Strategic Guidance and Recommendations*. Strasbourg: European Science Foundation.

Vermeulen, N. 2009. *Supersizing Science: On Building Large-scale Research Projects in Biology*. Maastricht: Maastricht University Press.

Webster, A. 2005. Social science and a post-genomic future: Alternative readings of genomic agency. *New Genetics and Society*, 24(2), 227-39.

Weinberg, A.M. 1967. *Reflections on Big Science.* Oxford: Pergamon Press.

Wiley, H.S. 2006. Systems Biology – Beyond the Buzz. *The Scientist*, June, 53-7.

Chapter 12

Genetic Discrimination 2.0: The Un/Differentiating Gene in Insurance

Ine Van Hoyweghen

Life, a constant process of innovation and creation, wild life that in certain respects defies domestication, suddenly appears in industry. (Callon 2007: 154)

With the advent of the Human Genome Project and the introduction of 'molecular medicine', genes have shown their appearance in new ways into society, as for example in the European insurance industry. From the outset there has been forceful debate on the use of genetic information in private insurance in many European countries. People feared that the use of genetic testing would render individuals uninsurable, leading to a 'genetic underclass'. Many concerned groups and policy makers have argued for solidarity for 'the genetic at risk' and called for regulatory action to prevent 'genetic discrimination'.

Genes have thus made their way in the insurance industry, as a peculiar bio-object. A new form of life is made the object of the biosciences, but how to manage it in European insurance markets? In taking up this question, this chapter's focus is on the multiple meanings of the bio-object as it circulates between different sectors of society. How did the bio-object of genes, associated with molecular medicine, find its way into the insurance assemblage with its processes and technologies of risk classification? To be sure, life has always been an object of knowing, representing and intervening in insurance. Francois Ewald (1986) long noted the significance of life insurance as a biopolitical technique of governance with the biological being as the object of value and governance. Life insurance has always traded statistically in the mortality and morbidity of the living creature through its actuarial practices, classification of population groups and risk selection. The legitimacy of insurance risk selection rests in the economic principles of adverse selection, moral hazard and the principle of actuarial discrimination, where people have to pay a premium according to the mortality risk they represent. To calculate what, actuarially, is 'a life', medical devices and actuarial tables have been developed to categorize and classify people into 'standard', 'substandard' or 'excluded' lives. What happens then when the object of life insurance – life itself – is being molecularized? In taking up this question, my concern is with the effects of the introduction of genes in the making of difference in insurance risk classifications. Life insurance has always been a 'machine of difference' *par excellence*. In sorting out people into 'standard' vs. 'non-standard' risks, insurance classification is a process of discrimination, of allocation of particularities by identifying differences (Bowker and Star 1999). As Stone (1993: 299) puts it: "Insurance underwriting, far from being a dry statistical exercise, is a political exercise in drawing the boundaries of community membership". How then are genes intervening in the politics of difference of European insurance markets? What are the consequences for the configuration of solidarity in insurance?

That way, the chapter matches a source of social dynamics that has received little attention until now: the connection between the biosciences and the proliferation of the social. Michel Callon for example argues how developments in the biosciences may contribute to the appearance of matters of concern which coincide with the emergence of new social groups or concerned groups ('orphan groups' and 'affected groups') (Callon 2007). To denote this mechanism of the emergence of new social relations and social divisions due to the biosciences, Callon proposes to use an expression coined by Strathern (1999: 174), the 'proliferation of the social.'[1] The proliferation of genes, proteins, embryonic cells, GMOs and so on, as 'living entities' (Caliskan and Callon 2010: 6) currently produced in laboratories through bioscientific practices may raise problems of categorization, standardization and framing in economic markets. This difficulty in 'taming'[2] the 'wild life' (Callon 2007: 154) of genes may result in political controversy and the proliferation of new social identities and relations. In this way, genes can be important operators of identity and solidarity, through the emergence of new identities and groups around genes and its political mobilization. Callon thereby analyses the dual role of genes, both as being configured into the collective, and in affording new social relations, that is, in performing the collective.

Building on the work of Callon and others in the field of STS,[3] I will trace in this chapter how genes have been integrated in the European insurance assemblage and how, at the same time, genes may perform the collective in new and unforeseen ways. In taking up these themes, I want to pursue the tension between the differentiation and dedifferentiation that genes introduce in the practice of insurance classification. I will explore how genes may be capable of dividing as well as linking people together in new ways in insurance (by proliferating and flattening difference). In exploring this game of un/differentiation, the aim of the chapter is to articulate *the transformative character* of bio-objects. In showing their 'wild' ontology, bio-objects may generate new divisions of the social and

---

1   As Strathern (1999) puts it, it is through the mediation of nonhumans (e.g. science and technology) that the social proliferates and human persons are produced. See also Callon and Law (1997).

2   The notion of 'taming' refers to the mutual adjustment of technology and the social and is related to the notion of 'domestication' (Latour 1987, Callon 1986). On the difficulty of 'taming' living/natural entities, see, e.g., Latour (2004).

3   I rely on those approaches in science and technology studies, which have taken up the question what the hybrid processes of categorization in biosciences, states and markets entail for the construction of identities and social relations (see, e.g., Epstein 2007, Rabinow 1996, Biehl 2004, Beck and Niewöhner 2009).

new configurations of solidarity, in challenging for example the modernist boundaries between the normal and the pathological in insurance (Rose 2001). Life insurance thus may furnish as a uniquely interesting intersection for studying the interplay between bio-objects and the social, by exploring the new identities and biosocial relations involved in the manufacture of biosciences, law and insurance classifications. The chapter builds methodologically and analytically on research on the issue of genetics in the 'zone' (Barry 2001) of European insurance.[4] A detailed analysis of case studies is beyond the scope of this chapter, as the chapter's purpose is mainly programmatic, fleshing out some important issues for discussion.

In my analysis, I distinguish two different moments in the process of 'taming' genes in the zone of European insurance and its respective re-configurations of solidarity. The first stage will look at the introduction of genes in insurance markets and the specific way this form of life has been made governable through the development of Genetic Non-Discrimination Acts (GNDAs) in European countries.[5] In this paradigm of 'genetic exceptionalism', solidarity is crafted towards the group of 'the genetic at risk' in insurance. Subsequently, the second stage examines how this 'new regime' of GNDAs-in-insurance is working in practice and explores its real-life consequences. It indicates some important intricacies of GNDAs in coming to terms with the 'wild life' of genes and it suggests how genes may re-configure solidarity in the zone of European insurance in new and unforeseen ways.

Genetic discrimination: The coming-into-being of GNDAs in European insurance

When the use of genetic technologies was fostered by the Human Genome Project in the nineties, one of the most contentious debates on its 'social impact' were to be found in the field of insurance markets. Should insurers be able to use genetic testing as a means to select out insurance candidates? This has been the thrust of debates in the public realm, policy makers and parliaments throughout Europe and has resulted in the establishment of Genetic Non-Discrimination Acts (GNDAs) in

---

4   I use Barry's notion of 'technological zone' to depict the hybrid spaces at work in the governance of European insurance, including a hybrid of national and transnational state and market actors. My analysis is based on written sources and interviews with key informants from the EU and EU country-based policy field, and on fieldwork research in insurance and reinsurance companies (see, e.g., Van Hoyweghen 2007; Van Hoyweghen and Horstman 2009).

5   The term 'Genetic Non-Discrimination Acts' (GNDAs) is coined by the author to denote the regulatory approaches – ranging from prohibitive to fair limits approaches – which have been applied to regulate the use of genetic testing in the field of private life and health insurance in European countries.

insurance.[6] GNDAs are a way of governing the issue of genes in insurance markets. However, it is not self evident that policy makers resort to state regulations to intervene in private markets. To understand the popularity and rise of GNDAs, it is necessary to get insight in the mechanisms that helped to define the issue from a problem affecting a small group ('the genetic at risk') into a major public issue, leading to the establishment of regulatory action in the field.

In general, what underpinned these debates has been a significant fear of 'genetic discrimination'. For one thing, underlying this fear has been a deterministic vision of genes as ultimate causal constituents. Both advocates and proponents in the debate used the Huntington's gene example, a rare monogenetic disorder gene, as the exemplary to underline their claims, framing the debate in a hyperbole (Van Hoyweghen 2007). At the same time, underlying the fear of 'genetic discrimination' have been preceding concerns and shared memories of eugenics (Lemke 2005). In addition, genetic testing was considered as one of the most intrusive examples of what third parties could do with personal data. But more importantly, underlying this fear has been a deep-rooted 'vision' (Ewald 1986) of insurance as a technology of solidarity, and not of discrimination. Clearly, twentieth-century European insurance arrangements have been identifiable by a deeply rooted "culture of solidarity" (Hinrichs 1995). In most European countries, access to insurance is seen as a social good, a right, not a privilege (Chuffart 1996). Within this vision, the issue of genetic testing in insurance was seen as an exemplary of the decline of the ethos of solidarity, expressing wider concerns of discrimination and social exclusion against the background of recent 'privatizations' of European welfare states.[7] Underlying the fear of the use of genetic testing in insurance has thus been a particular alignment of genetic and economic determinism, captured in the trope of 'genetic discrimination'. Discourse about 'genetic discrimination' in Europe has been wedded to narratives of genetic determinism, eugenics and a European insistence to social inclusion and non-discrimination. Within this framing, 'genetic discrimination' became defined as an issue requiring *special* treatment in the policy arena. This is how the issue of 'genetic discrimination' got aligned within

---

6   It is estimated that over 30 European countries have limited the use of genetic information in insurance underwriting through a combination of national and transnational legal instruments (Yoly et al. 2010).

7   European welfare states have a particular insurance 'imaginary' (Ewald 1986) in their view of the social role of insurance. Most European welfare states provide some form of universal access to health care and private health insurance has not played a significant role in the European Union (EU). Yet, current health care reforms in Europe towards the 'privatization' of health care have generated concerns of increasing social exclusion and the rise of a 'two-tier health system'. At the same time, the particular historical constellation of private/social boundaries in European welfare states has produced a social value vested in private insurance, reflected for example in broad risk pooling and lenient underwriting (typified as the 'Continental Underwriting' approach). Genetic testing – as a new tool to select 'the cream of the crop' – has raised concerns of increasing risk stratification in European private insurance.

a paradigm of 'genetic exceptionalism', where genetic information is believed to differ from other medical information, and therefore to be treated in a special way.[8]

To some extent, genetic patient groups have been active in mobilizing this view of 'genetic discrimination' as special and therefore requiring specific regulatory action. Genetic patient groups, such as the Genetic Interest Group (UK) or the Dutch Genetic Alliance (VSOP), have protested actively against 'genetic discrimination', through media attention, political campaigns and lobbying. As 'affected groups' (Callon 2007), they have used the gene as a device for mobilizing their specific concerns, while demanding compensation for their special status in insurance. These groups have also formed important partnerships with existing 'orphan groups' (Callon 2007) of insurance, such as patient organizations that are usually excluded from the insurance framing. But more importantly, 'genetic discrimination' has enabled alliances with other – originally non-orphan – publics and state actors, in making them concerned about genetics as an exemplary case for wider concerns of social exclusion. Idioms like 'the genetic underclass' have strongly appealed to a European welfare state ethos of solidarity. In some European countries, politicians and other state actors have been so captured by the idea of social inclusion that they have taken immediate policy action to restrict genetic testing in insurance. At the same time, 'genetic discrimination' has found alliances with medical professionals and with EU innovation policy actors active in the EU 'Knowledge Based Bio-Economy' because of concerns that people would defer from genetic testing (and thus thwarting genomic innovation) by the fear of 'genetic discrimination'. From another front, 'genetic discrimination' has aligned with concerns of human rights organizations, having developed in important transnational normative documents on genetic non-discrimination. In Europe, the Council of Europe (CoE)'s 'Oviedo Convention of Human Rights and Biomedicine' (1997) has clearly set the tone by prohibiting any form of discrimination against a person on grounds of his or her genetic heritage and restricting the use of genetic tests to health purposes or scientific research.[9] This 'montée en généralité' (Boltanski and Thévenot 1991) of the concerns of 'the genetic at risk' has been very effective in making 'genetic discrimination' an important societal issue that needed specific regulative attention.[10] But the action

---

8   See, e.g., Murray (1997) who argued against this genetic exceptionalism.

·   9   The Council of Europe (CoE) was set up to promote democracy and protect human rights and the rule of law in Europe. The Oviedo Convention on Human Rights and Biomedicine 1997 is legally binding for those members of the European Community that have ratified it. As of February 2011, 26 European states have ratified it. Retrieved: 10 February 2011 from: http://conventions.coe.int/Treaty/Commun/ChercheSig.asp?NT=164 &CM=7&DF=10/02/2011&CL=ENG

10   Including the idea that 'genetic discrimination' was made into an issue of 'state interest' as well. The above makes apparent the active role of national and transnational *state actors* as 'concerned groups' in having mobilized the gene as operators of identity and solidarity – with the effect of safeguarding ('welfare state solidarity') and/or engineering ('EU Bio-Economy') a national/European identity.

of the gene seems to go further than this. That is, 'genetic discrimination' not only stimulated the concerns of others for the problem of 'the genetic at risk' but was translated as well into an issue 'affecting all of us', by referring to the idea that 'we all have genes, so discrimination could happen to all of us.' As the director of the UK Genetic Interest Group said: "The fact that we are all susceptible to genetic diseases is a strong factor for solidarity".[11] In this way, 'we all', as potential genetic risks, have become 'affected groups' (Callon 2007) in the genetics and insurance issue. Genes may perform here as an important operator of solidarity, in linking the 'genetic at risk' with humankind.

This work of collective integration that genes have enabled in linking people together has given way to a *hybrid coalition* for political action in the regulatory field and for a common desirability for genetics-specific legislative action. To some extent, the concerned groups, struck by 'genetic discrimination', have ranged from the 'genetic at risk' to the whole collective. The alignment of different groups and concerns may explain why GNDAs have been so successful, as there started to be "a trend to legislate" in Europe (Comité Européen des Assurances [CEA] 2000: 2).

> But if you go to the governments, even though the national context is completely different, the first thing policy makers do is to look on the internet and see what other governments are doing. They really do that. And they just pick whatever . . . they don't care what the insurance system is in that country. They just go and say: 'Oh that must be a solution then, because we're not going to generate a new solution or something.' [. . .] Particularly in the number of European countries that in the early '90s has decided to legally ban genetics in insurance. That has generally set some kind of standard. (Interview with chief underwriter, reinsurance company B, September 2005)

Apparently, GNDAs have become the gold standard to make genes governable in the zone of European insurance. With the installation of GNDAs, state administrators have installed and cemented the new category of 'genetic' into insurance regulations. Genes have been 'tamed' in the zone of European insurance by an alignment of genetic and economic determinism, captured in the trope of 'genetic discrimination'. In this new framing, solidarity is expressed to the specific category of 'the genetic at risk'. In this way, 'genetic discrimination' has been made an *exception* in the politics of European insurance markets.

Genetic discrimination 2.0: European GNDA strategies put to the test

With the installation of GNDAs, we can begin to explore and map their effects in the zone of European insurance. While these governance technologies are

---

11   Retrieved 12 February 2010 from: http://ec.europa.eu/research/biosociety/news_ events/news_genes_medicalcare_en.htm

developed to solve the issue of genes in insurance, it increasingly becomes clear how the workings of GNDAs may be more complex than expected.

In fieldwork research in Belgian insurance companies, I documented how, despite a national legal ban on the use of genetic information in private insurance (LVO 1992), medical underwriters still used genetic information, interpreting the legal definitions of 'genetic information' in multiple ways. At the same time, insurers had to fit the new legal category of 'genetic' into already existing insurance risk classifications – family history, risk factors associated with lifestyle choices and other forms of testing (e.g. blood testing, cholesterol testing). Insurers have been using family history, as well as notions of inheritance, since the very beginnings of the industry, and these notions conflicted with the new definitions of the category of 'genetic' in Belgian law. Medical advisors indicated how delicate it was to filter 'genetic information' from the medical files they received from GPs in order to assess the applicant's insurance risk (Van Hoyweghen 2007).

At the same time, new developments in genomics may complicate the workings of GNDAs in insurance practice. One of the consequences of the shift from genetics (i.e. gene-oriented approaches) to genomics (i.e. genome-oriented or systems approaches) is that the distinction between genetic and non-genetic disorders has become an artifact. The rise of the complexity paradigm in genomics may conflict with the reductionist legal definitions of 'genetic' in GNDAs. This is for example reflected in the abundant legal commentary on GNDAs arguing how these laws' definitions seem to be outdated, as the original scope of protection provided by most GNDAs has been extremely narrow. For example, some GNDAs exclude chemical tests, blood tests and routine laboratory tests from the definition of genetic tests. Considering that all genetic tests are chemical tests, many are blood tests and an increasing number are routine, it is argued that many GNDAs in insurance provide only the illusion of protection. In insurance practice, medical advisors have been using more indirect forms of genetic testing, such as genetic information derived from chromosomes, proteins or via routine urine or blood tests and there is increasing debate on where to draw the line.

'Taming the wild life of genes' in insurance by GNDAs may thus be more complicated than expected. To some extent, genes may begin to 'strike back' (Latour 2000) to the paradigm of genetic exceptionalism. Increasingly, policy makers, patient groups and insurers begin to understand that 'the genetic' has become a problematic category. Below I will explore some possible *routes* to indicate how the co-production of genes, GNDAs and insurance may give way to the emergence of new concerned groups and new social divisions in European insurance.

GNDAs and the rise of new orphan groups: Challenging the 'right to underwrite'

In the paradigm of genetic exceptionalism, people with genetic risks are considered the potential 'victims' in insurance. If we shift attention to the workings of GNDAs in insurance practice, GNDAs may come with unanticipated consequences, such as increasing discrimination of other groups in insurance.

In Belgian insurance practice, the interplay of insurance risk classifications and the legal GNDA categorizations has resulted in an unanticipated re-categorization of morals in insurance practice (Van Hoyweghen 2007). While the Belgian law framed th(ose at) genetic risk as 'victims' in insurance, it reinforced the idea that lifestyle risks, like smoking and overweight, are a matter of individual responsibility. In insurance practice, this framing has resulted in an increasing ascription of responsibilities for lifestyle risks and increasing discrimination. Smokers, the obese and people who do not comply with prescribed therapy are confronted with sharply increased prices. By giving exclusive legal protection to the group of 'the genetic at risk', other risk groups are unintended being underprotected in insurance. This attention for the new gaps that GNDAs create is currently discussed as well in insurance practice in regard to other forms of 'predictive testing'.

> We have the silly situation here now that, depending on how people test for a disorder, we can use this information or not. It doesn't make sense. Cystic fibrosis is a wonderful example. You have a couple of good markers for CF, the sweat test for example. And you can do the same by genetic screening. And now with the installation of these gene-specific laws, all of a sudden, if you do a sweat test, you will be rated but if you do a genetic test you wouldn't, because we are not allowed to use it. I mean, what kind of disadvantage do you give to the people with the sweat test? (Interview with Head medical underwriting department Reinsurance Company C, June 2005).

Also in legal commentaries it is discussed that GNDAs do not protect people with other characteristics—such as a finding on a colonoscopy—that predispose them to illness, though they, too, are considered in insurance more expensive to insure than the average person. The successful attention to the specificities of the group of 'the genetic at risk' in GNDAs may produce the emergence of new 'orphan groups', who realize they are excluded from the exclusive protection of GNDAs. These groups may compare themselves with the GNDA category of 'genetic' and argue that they are treated unfairly before the law.

This 'game of comparison' of new orphan groups with the protected group of GNDAs has been recently materialized in European regulatory documents with the introduction of the concept of 'predictive testing' in extending the protection of genetic testing in insurance to other forms of predictive testing. The concept turned up in a document of the German Nationaler Ethik Rat on predictive health information in the context of insurance contracts (NER 2007). Currently, it is appearing on a European policy level as well, in the Council of Europe's (CoE) preparations of a legal instrument on the issue. Where the CoE consultation rounds started with the topic of 'genetic testing' in insurance, the topic is now likely to be extended to consider 'other medical examinations providing predictive health information' in European insurance.

The institutionalization of GNDAs may stimulate mobilization for more inclusive non-discrimination legislation in European insurance and contestation on the general insurance principles of discrimination. Recently, European regulators and governments have been increasingly restricting the insurance industry's access to medical information by installing non-discrimination regulations.[12] According to the reinsurance company Swiss Re:

> ... sensitive issues such as HIV and genetics have been key catalysts in turning discrimination, privacy and entitlement into key regulatory issues for insurers. (Swiss Re 2007: 6)

The introduction of genes in insurance markets, by producing affected groups and protection policies, may make other groups (human rights organizations, disability organizations) and state actors (appealed by an ethos of welfare state solidarity) increasingly aware and concerned about the general insurance framing and the normal 'victims' of this framing. This may result in claims for more inclusive non-discrimination rights in insurance, challenging the "right to underwrite" in European insurance (Swiss Re 2007: 5). Instead of solving the issue of 'genetic discrimination' with GNDAs, the issue of 'genetic discrimination' may extend towards challenging the principle of medical risk selection in European insurance. Genes may be a powerful operator of solidarity here in turning 'genetic discrimination' into a more general concern of 'discrimination' in insurance.

'All individuals are orphans': The splintering of the GNDA group

By enacting GNDAs in insurance, the intention has been to provide specific protection for the GNDA group of 'the genetic at risk'. In trying to fix the GNDA legal category of 'genetic' with the insurance categories of 'standard' vs. 'non-standard' risks in insurance practice, new contestation may however arise with increasing social comparison and competition *within* the GNDA group of 'the genetic at risk'.

This is for example what happened in the debate in the Netherlands on the insurability of people with Familiar Hypercholesterolaemia (FH) (see also Merkx 2007). Despite the introduction of a moratorium on the use of genetic information in insurance and the enacting of the Medical Examination Act (WMK 1997), the issue of genetics and insurance re-kindled a heated public debate in 2000. Life insurance companies were accused to violate the law by excluding people with familiar hypercholesterolaemia (FH) from life insurance. The debate was held in different places, for example in the Parliament, the Dutch Health Council, the Ministry of Health and in self-regulatory platforms. In these debates, the category of

---

12 See, e.g. the Council of European Union 'Gender Directive' (Council Directive 2004/113/EC) and the 2008 draft of the Council of European Union Directive 'prohibiting discrimination based on religion, disability, age or sexual disorientation' restricting private insurer's underwriting process (Swiss Re 2007).

'genetic' of the Dutch law was contested and discussion raised on how to categorize FH patients in regard to insurance classifications ('standard vs. nonstandard risks'), legal classifications ('genetic vs. non-genetic') and the multifactorial character of FH. Various stakeholders started to make comparisons of how the FH group differed from other groups based on specific characteristics and how this group was fairly/ unfairly treated in regard to insurance and/or legal classifications. Typical statements were, for example: 'our FH risk is the same as that from people who have a standard mortality risk', 'our FH risk is lower than the risk of diabetes patients', and 'it would be unfair if people with FH would be protected, while people with high cholesterol would have to pay.' In due course, who was to count as 'an FH patient' and what was to count as 'genetic' changed considerably and various social distinctions emerged: FH versus other genetic risks, genetic versus non-genetic risks, treatable versus non-treatable diseases, genetic versus non-genetic cholesterol risks, 'good' versus 'bad' lifestyle FH patients, and so on. Each time, a different aspect of the FH patient's health was highlighted in these comparisons. By emphasizing the differences in regard to certain groups, an image was created of FH patients as a separate category. By, at the same time, stressing the similarities with other groups, which were already receiving special protection in insurance, the image of FH patients was placed on an equal footing with those of other groups. This 'game of comparison' finally resulted in an increasing fragmentation of 'the FH group'. Typical statements were, for example: 'only those FH patients who are not compliant with the GP advice have a higher risk', 'those with a bad lifestyle get more frequent medical problems'. Because of the competition of claims for equality before the law and insurance classifications, the group of ' the FH patients' was split, splinted and re-categorized to such an extent that FH patients finally did not get specific protective policy actions in insurance, but were individually left out instead.

The above example indicates the rise of claims for a *differentiating* approach *within* the group of 'the genetic at risk' in insurance. With the appearance of genes as multifactorial, the legal category of 'genetic' may become contested. People 'made up' by the GNDA classifications may contest being categorized in the group of 'the genetic at risk', while preferring to be categorized along their personalized mortality risk (e.g. genetic susceptibility in combination with lifestyle risk). This may result in a proliferation of difference between individuals, generating a proliferation of claims for solidarity in the political arena, which may result in the splintering of the GNDA group of 'the genetic at risk'.

'We are all affected': Challenging the right to underwrite

By enacting GNDAs in insurance, the intention has been to install the policy choice of the *exceptional status* of genes in insurance markets. One condition of the emergence of GNDAs has been the claim that 'genes affect us all'. However, this very same claim may well be translated again into an expansion of GNDAs towards the prohibition of all medical information in insurance.

The developments in genomics towards a multifactorial approach may stimulate debates over the definition of 'genetic' in GNDAs. Now that all medical disorders are understood as having a genetic component, there has been concern in the insurance industry that the 'older' definitions of GNDAs will be applied *all* medical conditions, in such a way that: "... Routine underwriting parameters such as family history, serum cholesterol and blood glucose, physical examinations, and virtually all common medical impairments will be banned" (Pokorski 1997: 108). If it is hard to distinct between genetic and non-genetic information, there is a tendency in EU governmental policy to extend the legislative ban to *all* medical underwriting. As this reinsurance spokesman said:

> And then, the solution what the politicians are doing now is that they of course realize that [blurring of boundary between genetic and non-genetic]. They say: 'Oh well ok, wait a second, that's a problem. Well then our solution is that we extend our definition in the laws. And include all the tests, all the medical tests.' And that's what we would like to avoid. Because there are really good medical tests that are allowed to be used in the insurance market. These are general tests that are used in a clinical setting. And now all of a sudden they should be considered genetic. We don't want to go there. I mean we end up with the x-ray is now being a genetic test! Then we have this pan solution on all insurance business on all medical tests that are available. I think no; that isn't where we want to go (Interview medical advisor Reinsurance Company A, April 2007).

These observations indicate another *route* of how the GNDA paradigm may be extended to claims for a more inclusive prohibition of medical underwriting in European insurance markets. Figure 12.1 shows how an insurance company analysed the regulatory trends of the European private life and health insurance market with respect to the issue of genetic testing.

GNDAs may be translated into a storyline of 'genetic data = medical data' (Figure 12.1), challenging again the 'right to underwrite' of the insurance industry. Because genes tend to 'flatten' differences, GNDAs may be sensible to extension and analogy-making. When all persons have genetic susceptibilities, all persons are potential 'non-standard' risks, and the discrimination issues that are at stake with 'the genetic at risk', may be translated into issues of concern for 'all of us'. Genes may be a powerful operator of solidarity here in flattening the difference between 'standard' and 'non-standard' risks by a normalizing discourse of genes (biological essentialism).

Conclusion

A first aim of this chapter has been to articulate how the bio-object of genes, as a new form of life in the biosciences, has been integrated in European insurance markets. When genes made their appearance in the insurance industry in the

Genetic Data: Private Insurance Regulation

Europe: Regulatory Trend

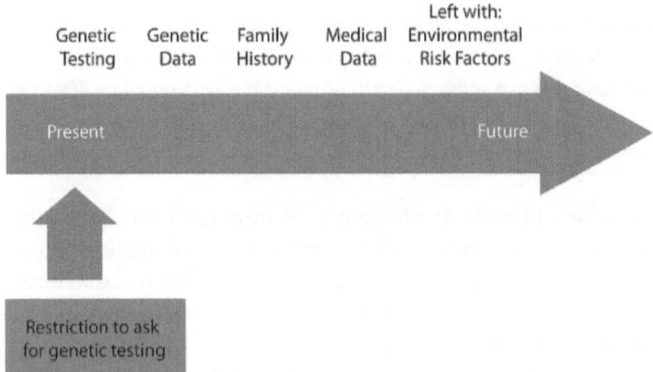

Figure 12.1   Medical genetics and regulatory considerations

Note: Reproduced with permission.

1990s, there was a shared concern that it was unfair to discriminate in insurance against 'the genetic at risk'. These matters of concern, underpinned by a paradigm of genetic exceptionalism, have aligned similar interests and resulted in a hybrid coalition of concerned groups, captured by the trope of 'genetic discrimination' and claiming for solidarity for the group of 'the genetic at risk' in insurance. By framing dilemmas of genetics and insurance as an *exception*, Genetic Non-Discrimination Acts (GNDAs) have been installed as a way of making genes governable in insurance markets. I further indicated the difficulties of the new state categorizations of GNDAs in 'taming the wild life of genes' in insurance practice. To some extent, genes may begin to resist the protective status they have been given by humans in GNDAs. This indicates the legal intricacies to deal with the multiple and diffracted character of bio-objects and the tension between legal form and biological agency. I then posited a possible proliferation of new concerned groups and state actors in European insurance. I explored how the interplay of genes, GNDAs and insurance markets may give rise to new controversies, resulting in the emergence of new concerned groups and new claims for solidarity in insurance. This means that issues of non-discrimination rights may be extended in European insurance by GNDAs. What was intended to balance out interests may create new social divisions (Strathern 1999). In telling this story, my concern has been to indicate the *transformative character* of bio-objects, in generating new identities and social relations. I have drawn attention to the role of genes in linking people together in new ways in insurance, as important operators of solidarity. By their capacity of flattening or proliferating difference in insurance, genes may be a 'catalyst' (Van Hoyweghen 2007: 160) for opening up the political in European insurance markets.

With the appearance of genes, the zone of European insurance has changed (even if genes are banned in insurance through GNDAs). The introduction of genes has managed to reconfigure the collective in which we, and/as genetic risks, live. This novel thing that genes bring into society may pave the way for a re-organization of European insurance markets, and more broadly, for reflection on the relationship between biosciences, markets and politics (Barry 2001; Callon 2007). By indicating how genes may be a catalyst in re-organizing the zone of European insurance, I have demonstrated the need for further research on the relationship between genes, insurance markets and the proliferation of the social. Sensibility to the co-shaping of genes and the social, and to the new identities and biosocial relations involved, I suggest, is essential for a better understanding of the politics of the zone of European insurance – including the engineering of a European identity and biosociality – , and for informed governance. We should not be naïve about the possibilities of classic governance technologies to deal with the biosciences, but we should have an open eye for the unruly, complex character of genes and the social. The proliferation of genes is likely to go together with a proliferation of the social.

References

Barry, A. 2001. *Political Machines: Governing a Technological Society*. London: Athlone Press.

Beck, S. and Niewöhner, J. 2009. Localizing genetic testing and screening in Cyprus and Germany: Contingencies, continuities, ordering effects and bio-cultural intimacy, in *The Handbook of Genetics and Society: Mapping the New Genomic Era*, edited by P. Atkinson, P. Glasner and M. Lock. London and New York: Routledge.

Biehl, J. 2004. The activist state: Global pharmaceuticals, AIDS, and citizenship in Brazil. *Social Text*, 80, 22(3), 105-32.

Boltanski, L. and Thévenot, L. 1991. *De la justification: Les économies de la grandeur*. Paris: Gallimard.

Bowker, G.C. and Star, S.L. 1999. *Sorting Things Out: Classification and its Consequences*. Cambridge, MA: MIT Press.

Caliskan, K. and Callon, M. 2010. Economization part 2: A research programme for the study of markets, *Economy and Society*, 39, 1, 1-32.

Callon, M. 1986. Some elements of a sociology of translation: Domestication of the scallops and the fishermen of St. Brieuc Bay, in *Power, Action, and Belief*, edited by J. Law. London: Routledge and Kegan Paul, 196-233.

Callon, M. 2007. An Essay on the Growing Contribution of Economic Markets to the Proliferation of the Social, *Theory Culture Society*, 24(7-8), 139-63.

Callon, M. and Law, J. 1997. After the Individual in Society: Lessons on Collectivity from Science, Technology and Society, *The Canadian Journal of Sociology*, 22(2), 165-82.

Callon, M. and Rabeharisoa, V. 2003. Research "in the wild" and the shaping of new social identities, *Technology in Society*, 25, 193-204.

Chuffart, A. 1996. Genetic testing in Europe. *Journal of Insurance Medicine*, 28, 2, 125-35.

Comité Européen des Assurances (CEA) 2000. *Some Helpful Guidelines for National Associations Confronted with Restrictive Legislation Regarding Insurance and Genetics*. Brussels: CEA.

Epstein, S. 2007. *Inclusion: The Politics of Difference in Medical Research*. Chicago, IL: University of Chicago Press.

Ewald, F. 1986. *L'Etat-Providence*. Paris: Bernard Grasset.

Hinrichs, K. 1995. The impact of German health insurance reforms on redistribution and the culture of solidarity. *Journal of Health Politics, Policy & Law*, 20(3), 653-87.

Joly, Y., Braker, M. and Le Huynh, M. 2010. Genetic discrimination in private insurance: Global perspectives, *New Genetics and Society*, 29(4), 351-68.

Latour, B. 1987. *Science in Action: How to Follow Scientists and Engineers Through Society*. Milton Keynes: Open University Press.

Latour, B. 2000. When things strike back: A possible contribution of 'science studies' to the social sciences, *British Journal of Sociology*, 51(1), 107-23.

Latour, B. 2004. *Politics of Nature: How to Bring the Sciences into Democracy*. Cambridge, MA: Harvard University Press.

Lemke, T. 2005. Beyond genetic discrimination: Problems and perspectives of a contested notion. *Genomics, Society and Policy*, 1(3), 22-40.

LVO (Wet van 25 juni op de Landverzekeringsovereenkomst) 1992. *Belgisch Staatsblad*, 20 August 1992, 18283-333.

Merkx, F. 2007. *Organizing Responsibilities for Novelties in Medical Genetics*, Twente University, PhD thesis.

Murray, T.H. 1997. Genetic Exceptionalism and 'Future Diaries': Is Genetic Information Different from Other Medical Information?, in *Genetic Secrets:*

Protecting Privacy and Confidentiality in the Genetic Era, edited by M.A. Rothstein. New Haven, CT: Yale University Press.

Nationaler Ethik Rat 2007. Predictive health information in the conclusion of insurance contracts [available at: http://www.ethikrat.org/_english/publications/Opinion_PHI_insurance.pdf [accessed 1 September 2010].

Pokorski, R.J. 1997. Medical underwriting in the genetics era: Selected transactions of the International Underwriting Congress 1997. *Journal of Insurance Medicine*, 29(2), 107-19.

Rabinow, P. 1996. *Essays on the Anthropology of Reason.* Princeton, NJ: Princeton University Press.

Rose, N. 2001. Normality and pathology in a biological age. *Outlines*, 1, 19-34.

Stone, D. 1993. The Struggle for the Soul of Health Insurance. *Journal of Health Politics, Policy and Law*, 18(2), 287-317.

Strathern, M. 1999. What is intellectual property after?, in *Actor Network Theory and After*, edited by J. Law and J. Hassard. Oxford, UK: Blackwell, 156-80.

Swiss Re 2007. Life risk selection at a fair price: reinforcing the actuarial basis. Zurich: Swiss Re Insurance Company [online] Available at: http://www.swissre.com/pws/research%20publications/risk%20and%20expertise/echnical%20publishing/life_risk_selection.html [accessed 9 October 2009].

Van Hoyweghen, I. 2007. *Risks in the Making: Travels in Life Insurance and Genetics.* Amsterdam: Amsterdam University Press.

Van Hoyweghen, I. and Horstman, K. 2009. Evidence-based underwriting in the Molecular age. The politics of reinsurance companies towards the genetics issue, *New Genetics & Society*, 28(4), 317-37.

Chapter 13

Still Life? Frozen Gametes, National Gene Banks and Re-configuration of Animality

Sakari Tamminen

It was Aristotle who once defined the essence of life. In his writings, such as *De anima* (On the Soul) and *De motu animalium* (On the Motion of Animals), he identified (animal) life as something that moves, that is animated by a certain vital force:

> Now determined before ... that the origin of other movements is that which moves itself, and that the origin of this is the unmoved, and that the first mover must of necessity be unmoved. (Aristotle 1986: 24)

The animating force, *psukhe*, was ultimately linked hierarchically to the prime mover itself always at rest. This is, of course, an idea later embraced by the neoplatonists and one lurking behind the well known metaphysics of the Great Chain of Being (Lovejoy 1964). But central to his argument about different forms of life – the multitude of plants, animals and humans as "life" manifests in and though particular living things – was that the key defining criteria of all life is movement. He coupled this explanation of movement with an explicit comparison to the orders of politics:

> We should consider the organization of an animal to resemble that of a city well-governed by laws. For once order is established in a city there is no need of a separate monarch to preside over every activity ... . In animals this same thing happens because of nature: specifically because each part of them, since they are so ordered, is naturally disposed to do its own task. There is, then, no need of soul in each part: it is in some governing origin of the body, and other parts live because they are naturally attached, and do their tasks because of nature. (Aristotle 1986: 54)

A number of other definitions for life (in its various forms) have been given over the millennia after Aristotle.[1] However, I think that his elegant metaphysics coupled

---

1 Thacker has recently re-articulated the question on "life itself" with relation to the Aristotle's "force of life" (*psukhe*) by claiming that there are three discourses embodying this concept in a different set of relations of meaning – the first being the conceptualisation

with an interesting metaphor linking animal movement and politics, organism and governance, nonhuman and human might prove useful in thinking about current technoscientific enterprises, especially those that go under the name of biodiversity calculations, material repositories of diversity such as nonhuman gene banks, politics and new economies based on global circulation of vital matters. Thinking with Aristotle's definition of animal life serves as a good heuristic guide to the current global biopolitics of nature related to these efforts – a politics aiming at maximising biodiversity by taking the animation of nonhuman life as its object of a number of interventions.

The argument presented in this chapter is twofold. First, one of the most prominent claims today is the informational character of genetics. Life has become calculable in its basic components, the base pairs found in the genome of any living organism provide an easy counting of variations of the "book of life" (Kay 2000) – this is the development that has been called the "molecularisation of life" (e.g. Rose 2007). This has allowed a number of biological fields resembling engineering – a collection of knowledge-practices going under the names of bio-informatics, post-genomics, synthetic biology – to test, develop and deploy new protocols, tools and practices to manage "life" for various purposes simultaneously re-articulating the truth of "life itself". However prominent these practices are in shaping the future of "life itself", the argument presented in this chapter runs in another direction as an invitation to look at another central technology of managing life today. It is a mundane and ubiquitous technology that has been mostly left unexamined concerning interesting questions they generate about the objectification of life.

Nonhuman life, plants and animals, are today witnessing a new reconfiguration of their corporeal existence. Today, much of nonhuman life, especially animal life, is going through a transformation from being animation to a cessation of this animation – as suspension of the very essence of its animality. Technically speaking this is a new modality of animal life managed in potentia, a virtual corporeality corresponding to a novel bio-object made possible by a number of

---

of this moving force as a "life spirit" or a divine spark, the second a biological understanding embracing a "quasi-vitalistic life force flowing through each living organism" and the third interprets the force in a psycho-cognitivist framework stressing the interpretation of perception and intellect (in an anthropocentric way). Accordingly, these three discourses roughly correlate to the understandings of "life itself" as found within the disciplines of theology, biology and psychology (Thacker 2010, 8-9). I will be articulating the question of life in this article in asking how – and in what way – could we understand the "life itself" with regards nonmoving, arrested corporealities that only contain the potentiality of movement. A reader familiar to Aristotle's arguments about the relation of life force, movement and potentiality will see interesting links with his here which, however, are not reducible to his initial understanding. The potentiality of movement the corporealities analysed in this article are able to generate are not reducible to the ones Aristotle thought as the sign or essence of living beings in that they transgress the bounds of natural philosophy and extend to the realm of (global) politics.

agrobiotechnological innovations. The roots of these reconfigurations go deep to the normalised technologies and practices of plant and animal breeding, the latter also serving as baseline models for current reproductive technologies such as IVF within human animals (Clarke 1998, Franklin and Ragonet 1998, Franklin and Lock 2003). The reconfigurations I am exploring here concern the motion and mobility of life taken as an object of knowledge within animal reproduction science and a corporeal target of related biotechnological interventions. Here, one of the most important technological interventions in objectifying life are cryopreservation practices that enable the intra-cellular arrestment of movement. These practices transform the actual living cells to cryopreserved bio-objects that generate new potentialities as potentially living, virtually reproductive corporeal beings.

Second, and at the political level, there are ongoing major changes that affect the movement of nonhuman genetic material at the global scale. For the last two decades, and more pronouncedly after the United Nation's Earth Summit in Rio de Janeiro in 1992, all forms of nonhuman life have become a renewed matter of concern (Latour 2004) for nation-states. Hailed by the ecological scientists as an international move to save the biodiversity of planet earth, the Convention on Biological Diversity (CBD, 1992) signed at the Rio meeting, however, also created an international political regime of genetic sovereignty to the signatory nations by granting them ownership to all native "genetic material of actual or potential value" (Article 15). New national bodies – now called national genetic resources – of nonhuman life emerged out of this global contract that overlay the boundaries of nation and its nature. In this new global politics of nature genetic resources are at once considered as part of the history of a nation as well as its naturally occurring nature as peculiar national-natural entities.

When nations gained the sovereign rights over their native patrimonial genetic materials, most of these were taken off from the global networks as a result of muddled ownership rights and compensation schemes (see Kloppenburg 1988, Brush 1999, Hayden 2003; cf. Helmreich 2009). The arrested life stored in national gene banks was largely removed from the international machinery of animal production for conservation purposes and for enacting a sedimentary politics of "nativity" securing potentially vital economics embodied in nations' nonhuman genetics. As a result, the extracellular motility of frozen native nonhuman materials was arrested too – the cellular level arrestment of movement was complemented by a global circulatory suspension disrupting the previous global biopolitics of nonhuman life.

In this chapter, I will start by questioning our understanding of life-as-animation – whether understood as cellular or extra-cellular, biology or politics. I will analyse the technical minutia of one of the most common way of "conserving life" and situate these in the context of a global politics of saving both biodiversity and nations' sovereign right to their national natural resources. This questioning on how animal life becomes an object of technical and political interventions, and how these transform the ideas "life itself" understood as animation, are exemplified in this article with an empirical example of gene banking work and related global politics. The short illustrations (identical to practices in elsewhere) given below are drawn from a four year ethnographic study within Finland's national animal genetic conservation programme between 2004-2008.

Cryopreservation – life as arrested animation

> The ability of life to survive the frozen state is at the very edge of our comprehension of animated processes. (Fuller, Lane and Benson 2004: i)

The dream of managing life by managing its movement in/with different temperatures, or by thermal regulation, has a long history. Systemising these dreams into empirical experiments Robert Boyle tried his ideas about the effects of cold to various substances and published a monograph on these already in 1665. Although his experiments were designed as "chemical experiments" (Christopoulou 2007), they were followed by other approaches to the problem of preserving life in cold temperatures during the eighteenth, nineteenth and twentieth-century. One of the most advanced areas of experimenting the effect of cold on living beings was the research on plant and animal reproduction to solve practical questions of agricultural production. Here, the most researched area was freezing gamete cells for breeding purposes (Leibo 2004).

Keeping with the trajectory of old techniques in sire hiring and its coupling of reproduction and economic interests, subsequent research concentrated heavily on extending the reproductive powers of male animals and, more specifically, to the male reproductive material: sperm. The twentieth century witnessed important advances in the management of reproductive processes, made principally through innovations in artificial insemination (AI) techniques. These steadily eliminated the need for the male animal to be physically present in all phases of animal reproduction. The fundamental problem in industrial animal reproduction was the short temporal window of vitality of the sperm outside the animal body – without any artificial technique, the sperm cells were dead in a matter of hours or minutes, depending on the surrounding conditions. The first AI techniques did away with the animal: fresh semen could be preserved for a few days in either egg yolk or milk-based extenders before fertility was lost and the cells died. In a complementary fashion, the development of sperm dilutants – appropriately known as 'extenders' – allowed more than one insemination to take place with one ejaculate. The initial volume could be divided into numerous batches of less vitally concentrated liquid mixes, which extended the reproductive force of a superior animal (Foote 2002, Salamon and Maxwell 1995).

As such, the technique of cryopreservation has a long history in the making for the use of animal breeding (Rutledge and Seidel 1983, Foote 2002), but a significant betterment, a "breakthrough", came in 1949 when Christoffer Polge (Polge, Smith and Parkes 1949) with his team in Cambridge could successfully freeze and thaw sperm without losing much of its vitality – a significant step for the field called cryobiology and animal breeding industry. The reconfiguration of the processes of reproduction quickly led to various innovations and to the rise of large-scale animal industry capable of a new level of industrialisation and capitalisation on animal (re)production (Foote 1981, Rasbech 1993, Clarke 1998), making it an internationally lucrative business.

Cryopreservation is a technique that gave the breeders a high level of control over the temporal and spatial dimensions of the reproductive powers of their animals. It did away with the constraints posed by cellular decay, literally by stopping the biological clock of the cell in extremely low temperatures. The novel potentiality of life that such an operation generates is twofold. What normally would indicate a potentially viable sperm batch – the amount of mobility (overall mobility within the sample volume) and motility of spermatozoids (individual movement) – is complemented with its reverse potentiality within the techniques of cryopreservation. The frozen vital material is viable only as far as all signs of life-as-mobility are gone and the cessation of animation within the sample is complete. Frozen life is a form of life at absolute rest, arrested life at the level of cellular processes.

In practice, however, not all gamete material is capable of sustaining life in that novel, quite artificial and hostile, laboratory ecology: the passage from in vivo to in vitro, from live animal to a banked genetic resource, is not one easily made by living matter. It imposes a novel kind of selection process for the sperm in comparison to the "natural" one. Active researchers on the field, such as Hiemstra, van der Lende, and Woelders (2006), for instance, note in their analysis of cryopreservation as a means of genetic conservation that '[t]here may be considerable differences between breeds and between males in the "freezability" of the semen. As a consequence, frozen semen of some genetically interesting breeds or males may not be suitable as a gene bank resource, or can be used only with a poor efficiency' (ibid: 47).

This "freezability" is itself a contingent and contestable category behind the selection process, one that operates on an inherently economic historical trajectory. Only the most efficient – biologically and economically viable – gametes are selected for the gene bank. The cryopreservation of sperm includes four basic steps.

Once the sperm is extracted from animals, it is put through a series of trials of strength – an often lethal process during which some sperm batches die and some survive. The preparation for the actual judgements is done by centrifuging the sperm in a more concentrated volume and removing much of the enveloping medium in which the sperm cells are naturally found. After this, a small drop is pipetted from the concentrated volume and placed under the microscope, with which a series of visual judgements of the vitality of the sperm are performed.

The visual judgement of the vital characteristics consists of three steps. The process starts with evaluation of the vitality of the spermatozoids by placing a small volume of sperm for an assay under the microscope and assessing the morphology of the sperm cells. Second, two interlinked dimensions of movement – with calculation of the relative quantity of mobile spermatozoids in the sperm and considering

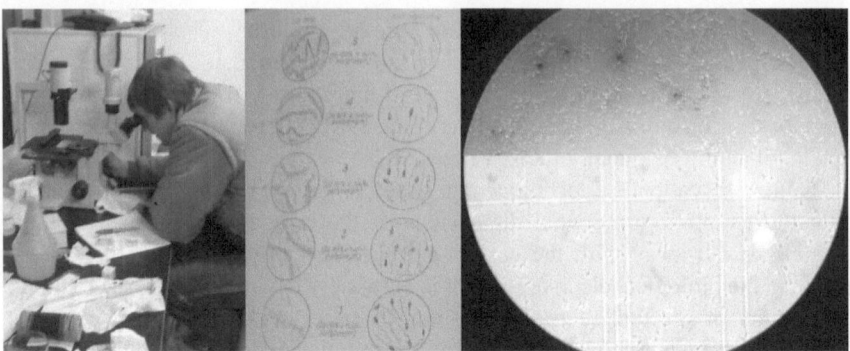

Figure 13.1   Performing visual judgements of mobility and motility

Note: From left to right, a) performing visual judgements of mobility and motility, b) the visual judgement standard, c) a picture of the object of vitality assessment with a normal slide on top and calculation of quality with the graticule slide on the bottom.

the direction of their movement – compose the overall visual judgement of their vitality. A certain concentration, level of viability (the normality of the structural morphology of individual gametes), overall motility, and direction of movement of spermatozoids are expected from the sample under evaluation. The concentration of spermatozoids within the volume is calculated by using visual translations of volume to area – the number of individual spermatozoids is calculated with the help of a special square engraved slide (see Figure 13.1).

A more detailed description of the process is needed in order to reveal how 'freezability' is constructed in action. If the first evaluation indicates that the overall quality of the sperm is good enough, it will then be 'extended'. The economic effectiveness of AI lies partly within this invention – from the early days, it was noted that the initial sperm ejaculate can be divided into smaller volumes without loss of too much of its reproductive vitality. Smaller volumes simply meant smaller volumes of sperm, and since normally the ejaculated volume contains a huge number of individual spermatozoids, dividing the volume did not pose too many problems for its reproductive powers.

The initial sperm amount is divided into smaller volumes and diluted with special 'extender' liquid. Extender liquids consist of a mixture of ingredients in which spermatozoids can survive (milk, egg yolk, and such), antibiotics for sterilising the mixture from bacteria, and cryoprotective agents (normally glycerine) that protect them from dying from a cold shock when frozen (Salamon and Maxwell 1995, Foote 2002). The 'extenders' are used, as the name aptly describes, to extend one batch of sperm into several batches and ready it for the actual freezing. With addition of the mixture of antibacterial ingredients and cryoprotective agents, one ejaculate can be extended to form many batches standardised in their volume and concentration of spermatozoids. Thus, the initial volume is diluted. The extended sperm are packed in small containers and sealed, making their new mode of being literally one of in vitro.

After counting, evaluation, and standardisation of the concentration of gamete cells, the small plastic containers (the ones regularly used are called 'Cassou straws') are immersed in the liquid nitrogen and frozen with carefully controlled lowering of the temperature until reaching minus 170 degrees Celsius. At this temperature, the cell 'clock' stops and the animation at cellular level is halted – the 'life' of the cells is suspended in a frozen state. The plastic straws are kept submerged for several minutes and then thawed slowly to revive the cell life. This is the last and the most important phase in the judgements leading to the construction of the notion of freezability for any given batch. Here, two aspects of animation of the spermatozoids are again crucial – motility and mobility.

This dual calculation procedure works by assigning the sperm batch under evaluation a discrete value from 1 to 5 and a symbolic notation consisting of plus signs (+). First, the motility of the thawed samples is calculated. If the sperm batch is visually judged to have up to 20 per cent motility of individual spermatozoids in its volume, it is assigned the number 1, if over 40 per cent it will be in category 2, and so on in 20 per cent steps, 5 being the best quality possible – displaying at least 80 per cent motility with a good direction (unidirectional mass) of overall movement. The second element contributing to 'freezability' is the direction of the movement, mobility, evaluated with pluses. If most of the sperm cells are moving in the same direction in what visually resembles waves or riverlike currents when seen through the microscope's lenses, the evaluative symbol assigned is five pluses (+++++), with the number of plus signs decreasing (to as few as one) the fewer waves the sample has.

With certain local thresholds, the figures for viability (morphological characteristics of sperm), the motility of the whole sperm population, and the forward mobility exhibited by individual spermatozoids are summed and may be judged 'freezable', and the sperm thus suitable and of good enough quality for ascension to the frozen repository. All of the standardised straws belonging to the 'unfreezable' batches are taken from the container and thrown away from the liquid nitrogen canisters. This is ultimately the last trial in the long series of visual judgements by which and where the 'normal' and the 'pathological' (Canguilhem 2007) of cryopreserved sperm are decided upon.

The collection, identification, standardisation, and banking create, craft, select, and process the animal and transform it from a biological being into its reproductive materials, with all the consequences that accompany this. The most important material consequence is that only certain genetic material embedded in the sperm is selected for the cryobank. Acceptance depends on gametes' resistance to the freezing protocols and alteration between the cessation of animation and resuming movement once thawed. Here, a novel technoscientific normalisation process (Foucault 1985) occurs, the end result of which is a new population of frozen sperm batches today called a 'gene bank'. In the end, this new frozen bio-object, quite literally, is cellular movement capable of being arrested – an embodiment of still life.

Globally generative relations – cellular circulation, genetic resources and the emergence of nonhuman nationhood

The possibility for suspension of cellular processes provided by cryopreservation techniques opened up not only a new temporal but also a new spatial scope. This yielded another kind of potential mobility, one that is extra-cellular. With the suspension of the biological processes, the sperm of superior animals could be transported virtually anywhere in the world to meet the great demand, as long as it was kept within an ecology – the unbroken 'cold chain' of cryopreservation – that provided the suspension of cell life in a frozen state. The frozen sperm changed from a local into a global commercial product in a very short time. It is simple-to-use and cheap technology for fast genetic enhancement and is easy to circulate around the world for various animals, such as cattle. Therefore, it soon became popular and in great demand, which also widened the spatial scope of the cryopreserved reproductive material enormously and created a novel global economy of frozen reproductive material (Cole and Cupps 1977, Brackett et al. 1981, King 1993, Foote 2002, Wilmot 2007)

At present, cryopreservation is still the world's most widely used biological technology in livestock-farming. Globally, over 100 million AIs in cattle, 40 million in pigs, 3.3 million in sheep, and 0.5 million in goats are performed annually. Of these, only about 4.5 per cent are performed with 'fresh' semen. The vast majority is cryostored, making AI and cryopreservation in most cases synonymous with each other (Thibier and Wagner 2002). For the last 20 years of the 20th century, more than 200 million frozen semen doses were produced worldwide every year. With these numbers, the theoretical size of the cattle sperm economy thus reached a value one could conservatively estimate at up to \$4-5 billion per year (of which about 5 per cent is international trade, see e.g. Gollin and Blackburn 2007).

As such, cryopreservation is part of the 'green revolution' in agriculture that from the 1960s onward rearranged nature with biotechnology striving toward genetic uniformity, toward greater control over, and capitalisation of, the vital processes of nonhuman life. This process and its methods of producing genetic monocultures have been so powerful that they have effected widespread concern about genetic erosion and the need for genetic conservation of both animals and plants (e.g. Shiva 1991). Out of this concern emerged new scientific disciplines aiming at the conservation of biodiversity, the pure difference found within nonhuman life. Thus, a paradoxical scientific relation to nonhuman life has been created, between its genetic destruction and salvation, in recent decades.

This ecological worry about genetic diversity has been linked to complex developments in international politics concerning nature and related questions of national sovereignty and nationhood that involve nonhuman forms of life.

The development of international nature politics has followed the paths of larger processes of globalisation, questioning the idea of national sovereignty and its powers to control the flow of capital, information, and – most importantly – corporeal forms of nonhuman life easily circulated such as cryopreserved gamete cells. 'Bioprospecting', the search for capitalisable forms of nonhuman life, as practised by the multinationals of the North within the territories of the 'biodiversity-rich' South has disrupted the latter nations' old ways of protecting their national interests, leading to accusations of systematic bioimperialism and biopiracy.

In 1992, the Convention on Biological Diversity (CBD) changed all that. Hailed in public as a landmark event in the global management of 'biodiversity', it brought unexpected and highly contested changes in international nature politics. In biology, 'biodiversity' is an umbrella term denoting differences found in nonhuman life, on scales as diverse as ecosystems, species, and populations. As a concept it first emerged in print in 1988 with the book Biodiversity, edited by the famous E.O. Wilson, composed of presentations given at a US conference called the National Forum on Biological Diversity. While the idea of, and research on, biological differences found in nature long pre-dates this conference, it is here that this difference was formulated simultaneously as a "global resource, to be indexed, used, and above all, preserved" (Wilson 1988: 3).

Within the CBD, however, 'biodiversity' performs mostly other functions than that embodied in the call to preservation. It is used as a common term for use amidst a number of problems of global ownership and issues of rights over nonhuman life. Here biodiversity is certainly no longer a common global resource but has become a contentious category of nature within a web of powerful international geopolitical interests and politics of nature. In many ways, the convention is, as Hayden (2003: 1) puts it, 'a living and much-contested document'. This is no surprise, as it is an international response to the outcry of the biodiversity-rich countries of the South about alleged biopiracy and new bioimperialism practised within their territories by multinational companies of the North. As a global convention, then, the CBD does not differ from others in the long list of global instruments aimed at securing rightful distribution between nations of the profits derived from natural resources.

However, it does stand out in one respect. In delegating rights and responsibilities related to various objects of nature, it is much more far-reaching in its implications than are other global contracts. With the CBD the previous (relatively) free global mobilisation and circulation of nonhuman life across national boundaries was put to an end. This resulted from three operational (re)definitions found within the convention. First, with Article 15 – the only internationally recognised hard-law part of the treaty – 'biodiversity' was effectively transformed from a biological understanding into something quite different. It became re-articulated through a molecular understanding of national nonhuman life with the concept of 'genetic resources', or as a collection of genetic material found in nature that could be turned into valuable resources – in actuality or potentially. Quite interestingly, the CBD remains decidedly ambiguous as to what counts as genetic resources. This ambiguous definition gives the signatory nations a possibility to claim sovereignty over nonhuman life of all kinds as long as they contain genetic material. The only condition is that this genetic material be native to – 'originating from' – the relevant signatory nation. As a result, the CBD potentially concerns all nonhuman life that has a (scientifically) proven 'country of origin', the condition set within the convention's text itself.

Second, 'biodiversity' became tightly enmeshed within particular political geographies of nation-states through this novel figure of genetic resources. By signing the convention, more than 150 nations (the US notably absent) decided that nonhuman life in all of its forms was to become nationalised and identified with a nationhood – they became objects of national genetic governance. Following the convention, all genetic materials exchanged (regardless of their use or the presence or absence of a compensation agreement) between nations ('parties to the convention') must have prior informed consent for the exchange and a certificate of origins for the materials.

Third, the signatory nations were bound to a new obligation to identify the totality of their genetic resources: they must provide an identified inventory of their patrimonial genetic materials found within the national territory. Previously, at issue had been individual cases of animals and plants, species, and other forms of nonhuman life, representing nonhuman nationhood. However, with the CBD, we see a novel imperative to calculate the totality of nations' nonhuman genetic material, across all of the traditional biological taxa (such as the plant and animal kingdoms). The convention obliges every signatory nation to produce a national inventory of their endemic genetic material (of value), regardless of its place within the taxonomic system applied in the life sciences. Since the convention, the nature of nationhood has been written through new corporealities – nations extend their reach to the molecular level of nonhuman life as nonhumans systematically become part of national communities. Nature becomes part of nation, and nation becomes part of nature as relations between nature and nation are generated. This is a rebirth of nations (Haraway 1997: 5) through technoscientific apparatus objectifying and ordering nonhuman life by gene-geopolitical interests and global politics found and ratified by the CBD.

Global biopolitics of arrested life

Today, biodiversity rich countries such as Brazil, Mexico and India have created specific regulations and national laws for the access and benefit sharing of genetic resources they consider native to their country. These also reflect in tight control over the export of nation's nonhuman genetic material. These explicit regulations have also generated novel relations between a number of parties – for example, between traditional communities and the representatives of the 'nation' or between academics and agricultural or pharmaceutical companies (Hayden 2003). The arrestment of genetic circulation have been noted also in international politics and taken into account, for example, in official assessments of other nations' capabilities to "cooperation". As the U.S. Department of State's Bureau of Western Hemisphere Affairs' entry on Brazil illustrates:

> Limitations to cooperation include substantial restrictions on foreign researchers collecting or studying biological materials, due to concerns over possible unauthorized taking and commercialization of genetic resources or traditional knowledge of indigenous communities (often referred to as "biopiracy").[2]

Many signatory countries, however, are still struggling in defining their regulations on access and benefit sharing and the question of ownership of frozen potential life becomes pronouncedly manifested in the context of these newly formed populations. For individual animal breeders and farmers the question is not only a theoretical pondering of the social contract and its constitutive rights of citizens to new kinds of natural properties. It is also a very practical question of business opportunities and economic rationalities. If the gene bank now practically defines 'genetic material' as cryopreserved sperm, how are the ownership rights and the capitalisation possibilities that go with it defined? If animal breeders today create frozen reproductive materials such as sperm from a valuable breed and start to sell it in worldwide markets, is this material also identified as containing genetic patrimony of the nation, thus making it the property of the nation instead of the farmers?

These kinds of ownership questions (and others that are just as problematic) remain vexing for the whole international community of signatories to the CBD (and also in the USA). When the signatory nations gained sovereign rights over their native patrimonial genetic materials, they were largely disconnected from the global networks, allowing only some distinct and separately specified gene banks that include cryopreserved materials to stay operational within the networks of trade and research (Whatmore 2002, Parry 2004). The variety in the many international and national regulations, combined with the right for every nation to devise its own regulatory frameworks has the result of muddled ownership rights and compensation schemes concerning genetic resources.

Simply put, the genetic sovereignty enacted with the CBD re-articulated national boundaries to the circulation of nonhuman life, thus demobilising them and banning international genetic transfer without due material transfer agreements (MTAs) between parties to the still-only-potential transaction. As a result, the intra-cellularly arrested life stored in gene banks of signatory countries to CBD after its ratification was removed from the international machinery of animal and plant production – either for national conservation purposes or for enacting economic sovereignty over valuable national resources.

Within this process, then, the extra-cellular motion of frozen nonhuman reproductive materials considered "native" was arrested too by national

---

2   Available at: http://www.state.gov/r/pa/ei/bgn/35640.htm

interventions to the neoliberal trade resulting from long-lasting agricultural and pharmaceutical interests in these materials. The 'pure' economic rationality of global agriculture and big pharma was interrupted by the ecological and, most importantly, by national interests in dealing with genetic resources. This is why frozen genetic materials such as animal gametes stored in gene banks, as particular material embodiments of national genetic resources, have become both intra-cellularly and extra-cellularly a still life.

Movement, life, politics: Still life and virtual animation

> But the "bio" of biopolitics remains this always-receding horizon – and this is a challenge for biopolitics itself … In a sense, then, the true object of biopolitics is not simply this or that form-of-life, but a notion of life itself that runs through the disciplined body and the secured population … From these touchstones we will derive a view of biopolitics as the attempt to govern life conceived as an anonymous, unhuman phenomenon of circulation, flux, and flow. (Thacker 2009: 135)

To note that animation and circulation have become an object through which life is today increasingly "managed" (in all of its senses) is not particularly new. Similar kinds of transformations – ones which are enacted through gene banking in global (human) tissue economies (Waldby and Mitchell 2006) – are well noted and questioned in the ongoing debates about "biovalue" and "biowealth" in the contemporary discourses of human biopolitics and governmentality.

Here, this transformation has been described as a transformation from an actuality to pure potentiality of life (following the Foucauldian tradition a large number of writers claim this, maybe the most recent sustained analysis can be found in Sunder Rajan 2006, a number of critical comments to this is provided by Agamben 1998; 2004 and Esposito 2008). However, I claim that within these discourses this transformation is mainly understood as a non-mattering potentiality-as-expectations, a capitalist form of creating virtual value to new forms of life through the stock markets and other volatile mechanisms of virtual capital.

As important as these economic mechanisms at large are, they do not suffice to explain how a number of rationalities, technologies and skills are needed to produce new bio-objects, new corporeal configurations of suspended life in flesh that are vital in generating new relations – both potential and actual. To complement these analyses, I suggest that two important questions concerning the status of "life itself" emerge when one looks at the political economies of nonhuman life that are intertwined in a way which urges to reassess the idea of "life" as target of economic (or other kinds of) interventions. This difficulty points towards some of the limits of our current understanding of both life and the politics that takes life as its object. First, when most of world's biodiversity (or nonhuman life in all its variety) is corporeally increasingly conserved in its reproductive potentiality in cryopreserved state rather as actual living biological beings, then how should we think of life and living after this corporeal transformation? Second, and given that biological life is increasingly transformed to its reproductive potentiality at cellular level, how is it that this virtual life is able to generate and actualise novel relations between nations and their natures in the sphere of nonhuman global biopolitics?

The questions point toward the problem of "life itself". Life at cellular level, it seems, is as much political as it is in the level of population control for national purposes. As Thacker puts it above, this is an effect of the new inter/national biopolitics aimed at arresting circulation, flux, motion of life through technoscientific apparatus of today. What emerges out of this process is a form of still life where the idea of "life itself" becomes re-articulated through the potentiality that intra- and extracellularly suspended life holds. The argument presented here is that the questions about life currently articulated through concepts of "bare life" (zoê) in its corporeal form and political ordering of life (bios) (see e.g. Agamben 1998, Rabinow 1999), become subsumed to each other. As such they cannot serve anymore as the only ways to understand "life", and, instead, new ways of questioning "life itself" must be generated.

At the corporeal level, the question about "life itself" becomes articulated by the gene-banked matter embodying complicated corporeal, economic and political relations described above. The material interests of different fields of reasoning – economy, ecology, and national – are best served by cryotechnology capable of arresting movement at the intracellular level. However, the same technology that generates the potentiality of biocapital by creating biovalue out of reproductive matter generates also a new articulation between the corporeal matter, of whatever nature, and the nation. Here, in the national genetic resource programmes within the order of the CBD, cryopreservation is used to arrest animation in order to maximise national biodiversity, translated as the pure national difference found in nature itself.

The arresting of motion at intracellular level, at the same time, actualises the potentiality of reproductive matter for extra-cellular circulation through global agricultural networks of animal production and biocapital. This potentiality is based on the promise of easy accumulation of added value for animal sperm via its circulation beyond territorial boundaries. However, it is precisely the ban of exports and extra-cellular circulation that also actualises the nation through its new nonhuman population found stored within the gene banks. The nation interpolates the reproductive matter and constitutes it as both ecologically native and economically viable 'genetic resources' through the mediation of cryoconservation technologies – technologies that themselves have a long genealogy of agricultural striving toward control and capitalisation of reproductive forces. Surprisingly, perhaps, at the level of material practices and corporeal substances, the generation of potentialities in the context of genetic resources leads to the revitalisation of the nation through a new life form that becomes a politically charged form of life – a still life.

I have claimed in this chapter that if life was once about movement, or self-generated animation by a force of life, and global biopolitics aiming at maximising the circulation of frozen corporealities embodying the potentiality of this movement, then now we have a suspension that of movement both at intra- and extra-cellular levels. From an animal life as actuality we have moved to its contingent national potentiality, a curious form of life that is virtual life in its corporeal form. And because of this suspension of movement at intra-and extra-cellular levels, genetic resources cannot anymore serve as bodies "inserted to the machinery" of production in ways which Foucault (1985) once envisioned being the modus operandi of biopower and related politics of life. This cessation of movement complicates both the Aristotelian conception of animal and Foucauldian conception of biopower. As such, cryopreserved vital national material, such as genetic resources described in the chapter, complicate two major threads of understanding and analyzing "life" today – as the animation of an organic object and as the object of political governance.

Thus, the general question I have tried to ask in the chapter is what is this kind of novel bio-object – what is still life? – and attempted to provide first re-articulations of the question about life in its objectified forms. One of the possible ways of answering the questions raised above points towards understanding life as not only as actual but also virtual animation (Deleuze 1997) capable of generating new movements in the relations between the animal and politics, organism and governance, nonhuman and human orders of life. What forces animate this virtual movement, and how exactly they are capable of generating this new form of animation, is a central question of our technoscientific contemporaneity.

References

Agamben, G. 1998. *Homo Sacer.* Stanford, CA: Stanford University Press.
Agamben, G. 2004. *The Open: Man and Animal.* Stanford, CA: Stanford University Press.
Aristotle. 1986. On the Motion of Animals, in *Aristotle's De Motu Animalium*, edited by M.C. Nussbaum. Princeton, NJ: Princeton University Press.
Brackett, B.G., Seidel, G.E. and Seidel, S.M. (eds), 1981. *New Technologies in Animal Breeding.* New York: Academic Press.
Brush, S.B. 1999. Bioprospecting the Public Domain. *Cultural Anthropology*, 14, 535-55.
Canguilhem, G. 2007. *The Normal and the Pathological.* New York: Zone Books.
Christopoulou, C. 2007. Robert Boyle's Experiments on Cold: A Study of the Role of Chemical Experiments. *The Proceedings of the 6th International Conference on the History of Chemistry*, 423-31.
Clarke, A. 1998. *Disciplining Reproduction: Modernity, American Life Sciences, and the Problems of Sex.* Berkeley, CA: University of California Press.

Cole, H.H. and Cupps, P.T. (eds) 1977. *Reproduction in Domestic Animals*. New York: Academic Press.

Deleuze, G. 1997. Immanence: A Life ... *Theory Culture Society*, 14, 3-7.

Esposito, R. 2008. *Bios: Biopolitics and Philosophy*. Minneapolis: University of Minnesota Press.

Foote, R.H. 1981. The Artificial Insemination Industry, in *New Technologies in Animal Breeding*, edited by B.C. Bracket, G.E. Seidel and S.M. Seidel. New York: Academic Press, 13-39.

Foote, R.H. 2002. The History of Artificial Insemination: Selected Notes and Notables. *Journal of Animal Science*, 80: 1-10.

Foucault, M. 1985. *History of Sexuality*. New York: Vintage Books.

Franklin, S. and Ragoné, H. (eds) 1998. *Reproducing Reproduction: Kinship, Power and Technological Innovation*. Philadelphia: University of Pennsylvania Press.

Franklin, S. and Lock, M. (eds) 2003. *Remaking Life and Death: Toward an Anthropology of the Biosciences*. Santa Fe, NM: School of American Research Press.

Fuller B.J., Lane, N. and Benson, E.E. 2004. Introduction, in *Life in the Frozen State*, edited by B.J. Fuller, N. Lane and Benson E.E. Boca Raton. CRC Press, i-vi.

Gollin, D. and Blackburn, H. 2007. *International Flows of Animal Genetic Resources: An Economic and Biological Analysis*. Available at: http://www.fao.org/AG/AGAInfo/programmes/en/genetics/documents/Interlaken/sidevent/5_1/Gollin.pdf [accessed 20 March 2011].

Haraway, D. 1997. *Modest_Witness @ Second Millennium.FemaleMan_Meets _ Oncomouse*. New York and London: Routledge.

Hayden, C. 2003. *When Nature Goes Public: The Making and Unmaking of Bioprospecting in Mexico*. Princeton, NJ: Princeton University Press.

Helmreich, S. 2009. *Alien Ocean: Anthropological Voyages in Microbial Seas*. Berkeley, CA: University of California Press.

Hiemstra, S.J., van der Lende, T. and Woelders, H. 2006. *The Potential of Cryopreservation and Reproductive Technologies for Animal Genetic Resources Conservation Strategies*. The Role of Biotechnology in Exploring and Protecting Agricultural Genetic Resources. Rome: Food and Agriculture Organization.

Kay, L.E. 2000. *Who Wrote the Book of Life?* Stanford, CA: Stanford University Press.

King, G.J. (ed.) 1993. *Reproduction in Domestic Animals*. World Animal Science Series, B9. Amsterdam: Elsevier Science.

Kloppenburg, J.R. 1988. *First the Seed: The Political Economy of Plant Biotechnology, 1492-2000*. Cambridge: Cambridge University Press.

Latour, B. 2004. Why Has Critique Run Out of Steam? *Critical Inquiry*, 30, 225-48.

Leibo, S.P. 2004. The Early History of Gamete Biology, in *Life in the Frozen State*, edited by B.J. Fuller, N. Lane and Benson E.E. Boca Raton. CRC Press, 347-70.

Lovejoy, A.O. 1964. *The Great Chain of Being: A Study of the History of an Idea*. Cambridge, MA: Harvard University Press.

Parry, B. 2004. *Trading the Genome.* New York: Columbia University Press.

Pistorius, R.J. 1997. *Scientists, Plants and Politics: A History of the Plant Genetic Resources Movement.* Rome: International Plant Genetic Resources Institute.

Polge, C., Smith, A.U. and Parkes, A.S. 1949. Revival of spermatozoa after vitrification and dehydration at low temperatures. *Nature,* 164, 666.

Rabinow, P. 1999. *French DNA: Trouble in Purgatory.* Chicago, IL: University of Chicago Press.

Rasbech, N.O. 1993. Artificial Insemination, in *Reproduction in Domesticated Animals,* edited by G.J. King. Amsterdam: Elsevier, 365-86.

Rose, N. 2007. *The Politics of Life Itself: Biomedicine, Power, and Subjectivity in the Twenty-First Century.* Princeton, NJ: Princeton University Press.

Rutledge, J.J. and Seidel, G.E. 1983. Genetic Engineering and Animal Production. *Journal of Animal Science,* 57, 265-72.

Salamon, S. and Maxwell, W.M.C. 1995. Frozen Storage of Ram Semen I. Processing, Freezing, Thawing and Fertility after Cervical Insemination. *Animal Reproduction Science,* 37, 185-249.

Shiva, V. 1991. *The Violence of Green Revolution: Third World Agriculture, Ecology and Politics.* New Delhi: Zed Press.

Sunder Rajan, K. 2006. *Biocapital: The Constitution of Postgenomic Life.* Durham, NC: Duke University Press.

Thacker, E. 2009. The Shadows of Atheology: Epidemics, Power and Life after Foucault. *Theory Culture Society,* 26, 134-52.

Thibier, M. and Wagner, H-M. 2002. World Statistics for Artificial Insemination in Cattle. *Livestock Production Science,* 74, 203-12.

Waldby, C. and Mitchell, R. 2006. *Tissue Economies: Blood, Organs, and Cell Lines in Late Capitalism.* Durham, NC: Duke University Press.

Whatmore, S. 2002. *Hybrid Geographies: Natures, Cultures, Spaces.* London: Sage Publishers.

Wilmot, S. 2007. From "Public Service" to Artificial Insemination: Animal Breeding Science and Reproductive Research in Early Twentieth-Century Britain. *Studies in History and Philosophy of Science Part C: Studies in History and Philosophy of Biological and Biomedical Sciences,* 38, 411-41.

Wilson, E.O. 1988. The Current State of Biodiversity, in *Biodiversity,* edited by E.O. Wilson. Washington, DC: National Academy Press, 1-18.

Index